Manchester minds

Manchester University Press

Manchester minds

A university history of ideas

Edited by Stuart Jones

MANCHESTER UNIVERSITY PRESS

Published by Manchester University Press
Oxford Road, Manchester, M13 9PL

www.manchesteruniversitypress.co.uk

British Library Cataloguing-in-Publication Data
A catalogue record for this book is available from the British Library

ISBN   978 1 5261 7632 5   hardback

First published 2024

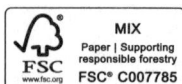

MIX
Paper | Supporting
responsible forestry
FSC
www.fsc.org   FSC® C007785

Typeset
by New Best-set Typesetters Ltd
Printed in Great Britain
by Bell and Bain Ltd, Glasgow

Contents

Contents

Contents

Illustrations

Every reasonable attempt has been made to obtain permission to reproduce copyright images. If any proper acknowledgement has not been made, copyright holders are invited to contact the editor via Manchester University Press.

Plates

Figures

Illustrations

Contributors

Katherine Ambler is Research Associate in History at the University of Northumbria. She completed a PhD on 'The Manchester Department of Social Anthropology 1949–1975' at King's College London in 2022.

Catherine Annabel is an independent researcher (PhD, University of Sheffield, 2022) and a contributor to *W. G. Sebald in Context*, ed. Uwe Schutte (Cambridge University Press, 2023).

Jonathan Aylen is Honorary Senior Research Fellow at the Alliance Manchester Business School, University of Manchester.

Joanna Bourke is Professor Emerita of History at Birkbeck, University of London, and author of *Birkbeck: 200 Years of Radical Learning for Working People* (Oxford University Press, 2023).

Matthew Cobb is Professor of Zoology at the University of Manchester. His most recent book is *The Genetic Age: Our Perilous Quest to Edit Life* (Profile Books, 2022).

Joe Earle is Executive Director of People's Economy, and co-author of *The Econocracy: The Perils of Leaving Economics to the Experts* (Manchester University Press, 2017).

Luke Georghiou is Deputy President and Deputy Vice-Chancellor of the University of Manchester and Professor of Science and

Technology Policy and Management. He is co-editor of *Mergers and Alliances in Higher Education* (Springer Open, 2015).

Elizabeth Gow is a Manuscript Curator and Archivist at the John Rylands Research Institute and Library, who recently completed a PhD entitled 'Enriqueta Rylands: The Public and Private Collecting of a Nonconformist Bibliophile, 1889–1908'.

David Hayton is Emeritus Professor of History at Queen's University Belfast. He is the author of *Conservative Revolutionary: The Lives of Lewis Namier* (Manchester University Press, 2019).

Sukhdev Johal is Professor of Accounting and Strategy at Queen Mary University of London, and co-author of *When Nothing Works* (Manchester University Press, 2023).

Stuart Jones (H. S. Jones) is Professor of Intellectual History at the University of Manchester, and is currently completing an intellectual biography of James Bryce for Princeton University Press.

John McAuliffe is Professor of Poetry at the University of Manchester and Associate Publisher at Manchester-based poetry publisher Carcanet Press. His most recent book is *Selected Poems* (The Gallery Press, 2021; Wake Forest University Press, 2022).

Chris Manias is Senior Lecturer in the History of Science & Technology at King's College London, and is the author of *The Age of Mammals: Nature, Development and Paleontology in the Long Nineteenth Century* (University of Pittsburgh Press, 2023).

Peter Dean Mohr is a retired consultant neurologist at Salford Royal Hospital, an honorary lecturer in medicine and advisor to the Museum of Medicine and Health, University of Manchester.

Peter J. T. Morris is honorary research fellow of University College London and co-author of *Henry Enfield Roscoe. The Campaigning Chemist* (Oxford University Press, 2024).

Mary Jo Nye is Professor Emerita of History at Oregon State University, and author of *Blackett: Physics, War, and Politics in the*

20th Century (Harvard University Press, 2004) and *Michael Polanyi and His Generation: Origins of the Social Construction of Science* (University of Chicago Press, 2011).

Peter Reed is an independent scholar in California, and co-author of *Henry Enfield Roscoe: The Campaigning Chemist* (Oxford University Press, 2024).

Derek Robbins is Emeritus Professor of International Social Theory at the University of East London. He is the author of *The Bourdieu Paradigm* (Manchester University Press, 2019) and editor of Tomoo Otaka, *Foundation of a Theory of Social Association, 1932* (Peter Lang, 2023).

Nancy Rothwell was President and Vice-Chancellor of the University of Manchester from 2010 to 2024, and the first woman to hold this role.

Gerardo Serra is Lecturer in Economic Cultures at the University of Manchester. He is currently working on his first book, under contract with Cambridge University Press, on economic knowledge and political imagination in Ghana.

James Sumner is Senior Lecturer in the History of Technology in the Centre for the History of Science, Technology and Medicine at the University of Manchester. He is currently developing a project on how non-expert audiences received and responded to computers and digital technology in twentieth-century Britain.

Karel Williams is Director of Foundational Economy Research Ltd, and co-author of *When Nothing Works* (Manchester University Press, 2023).

Rachael Wiseman is Reader in Philosophy at the University of Liverpool and co-author with Clare Mac Cumhaill of *Metaphysical Animals: How Four Women brought Philosophy back to Life* (Chatto and Windus, 2022).

Michael Worboys is Emeritus Professor in History of Science, Technology and Medicine at the University of Manchester and author of *Doggy People: The Victorians Who Made the Modern Dog* (Manchester University Press, 2023).

Natalie Zacek in Senior Lecturer in American Studies at the University of Manchester, and the author of *Thoroughbred Nation: Making America at the Racetrack, 1793–1900* (Louisiana State University Press, 2024).

Acknowledgements

Historians' acknowledgements usually begin by saying how long the book was in the making. This one, by contrast, went from conception to delivery of the manuscript in fifteen months, and I thank the many contributors for keeping to time to ensure that the book can appear on schedule in 2024. The project was formed in discussion with Luke Georghiou, Simon Ross and Emma Brennan, and with members of the University of Manchester's Research Group on University Histories, and I am grateful to them for their help. Emma, the book's editor at MUP, has gone far beyond the call of duty in ensuring the success of the project. Kate Hawkins provided invaluable help with the selection of images, the preparation of the manuscript, and much else.

The contributors are grateful to the many librarians and archivists who have helped locate materials. Material held by the John Rylands Research Institute and Library (JRL) is cited courtesy of the University of Manchester. Holders of rights in respect of the many images included are acknowledged separately.

The book is published as part of the University of Manchester's bicentennial celebrations. I am grateful to the University for the invitation to edit this book and for its support in enabling its completion. This is not, however, an official history, and contributors voice their own opinions, and not those of the University.

HSJ
March 2024

Abbreviations

BJRL	*Bulletin of the John Rylands Library* (or *Bulletin of the John Rylands University Library of Manchester*)
JRL	John Rylands Library
ODNB	*Oxford Dictionary of National Biography* (Oxford: Oxford University Press, 2004, http://www.oxforddnb.com/)
PP [British]	Parliamentary Papers
UML	University of Manchester Archive held in the University of Manchester Library
VCA	Vice-Chancellor's Archive (part of the University of Manchester Archive)

Introduction

Stuart Jones

Est. 1824

In marking its bicentenary in 2024, the University of Manchester asserts its descent from not one but two institutions founded in 1824: the Manchester Mechanics' Institution and the Pine Street School of Medicine. Nobody in Manchester realised in 1824 that they had established a university. The idea was quite conceivable, though idiosyncratic: a university for Manchester was proposed first in 1641, then again in 1783, and again in 1836, but not in 1824.[1] The Mechanics' Institution and the Pine Street School did not look like embryonic universities, and the journey from there to the University of Manchester of today was circuitous. In fact higher education in Manchester stretches back further than 1824, since Manchester Academy, where John Dalton taught, was founded there in 1786, and still exists today as Harris Manchester College, Oxford. Nevertheless, the Mechanics' Institution, and the Pine Street School too, were important landmarks in Manchester's educational, cultural and intellectual development. The bicentenary offers an appropriate time to launch a book that supplies, not a new institutional history, but a critical appraisal of the intellectual life of the University and its precursors.

The mechanics' institutes proliferated in the second quarter of the nineteenth century: by mid-century, there were some 700 in

1

the British Isles. In our opening chapter, Joanna Bourke captures the spirit that motivated their creators: the aim of promoting self-improvement through 'useful knowledge' among artisans and skilled workers. These aims were pursued in different ways in different places, and with varied degrees of success. There was no central organisation, and though each institute no doubt learnt something from its counterparts in neighbouring towns, they were much more profoundly shaped by their own urban environment. In Manchester, both institutions established in 1824 formed part of a wider network of knowledge institutions in the town, which included the Portico Library (1802–06), the Royal Manchester Institution (1823–24), the Manchester Statistical Society (1833) and the Manchester Athenaeum (1835).[2] Like these, the Mechanics' Institution was membership-based, although it operated a distinction between the 'honorary' members (middle-class patrons) and the 'ordinary' members (the students, as they were not generally called at the time). These bodies all had strong links with the Manchester Literary and Philosophical Society (the 'Lit and Phil'), founded in 1781. The surgeon Thomas Turner (plate 1), founder of the Pine Street School, was an active member of the Lit and Phil, where medical men were numerous: he lectured there on anatomy and physiology, as Peter Roget, of the *Thesaurus*, had done before him, and it was the success of these lectures in the autumn of 1822 that persuaded him to set up his School of Medicine. Benjamin Heywood, the first chairman and then president of the Mechanics' Institution, was treasurer of the Lit and Phil from 1815 to 1850, and Lit and Phil members were well represented among the Institution's other founders.

Two other important connections should be mentioned here. One was Cross Street Chapel, a mighty stronghold of Unitarianism, a denomination defined principally by 'rational Dissent' and a commitment to freedom of inquiry in religious and other matters. The chapel's members had been largely responsible for the establishment of the Lit and Phil, which held its first meetings there, and stalwart members later included William Fairbairn and James Heywood (Benjamin's nephew), leading figures in both the Lit and Phil and the Mechanics' Institution. Then there was the *Manchester Guardian*, founded in 1821 a few yards away from the Cross Street Chapel, to press the case for parliamentary reform in the wake of the Peterloo Massacre of 1819. Its first editor, John Edward Taylor, was a prominent member of the chapel, and a trustee from 1840. The personnel of these bodies overlapped to a remarkable extent, and the politics and knowledge culture of late Hanoverian and early Victorian Manchester cannot be understood without a grasp of the vitality of these networks.

The Lit and Phil and the *Guardian*, like the University, have recently investigated their founders' connections with the transatlantic slave trade and slave-grown raw materials. The depth of the British economy's ties with the slave economy in its widest sense are now well known. If we distinguish three strands in Britain's connections with slavery – investment in the slave trade, investment in plantations, and trading dependence on slave-grown produce – then it is clear that Manchester's connections mostly but certainly not exclusively took the third form. Natalie Zacek's contribution to this volume also draws attention to the Heywood family's historically close involvement in financing slave voyages. Manchester cotton manufactures were exported to Africa in large quantities and this trade formed part of the business model of slave-trading vessels. Some prominent local families, such as the Gregs of Quarry Bank, invested heavily in sugar plantations. That said, by the 1820s many of the founders of the Mechanics' Institution were supporters of the abolitionist cause, including both Benjamin Heywood and Robert Hyde Greg.[3]

1824 is by no means the only foundation moment the current University might commemorate. A more conventional date might be 1851, when Owens College was founded with the backing of a substantial endowment from the will of John Owens, a wealthy cotton merchant. Owens College, though not yet a university, was intended from the outset to provide an education akin to that offered in the universities. But as I suggest in Chapter 3, the transformative moment in the college's history came in 1870–73, when, with the support of a strikingly successful fundraising campaign among Manchester's business and professional communities, the college was re-endowed, relocated to more imposing premises on Oxford Road, and reconceived as a college that would genuinely belong to the city: a civic college, if not yet a civic university (see plate 3). The 'extension' of Owens College came at a strikingly important moment in university history, especially in the English-speaking world: in the wake of the Civil War, American universities as we know them today started to take shape: key milestones included the creation of Cornell University under Andrew Dickson White (1865), the appointment of Charles W. Eliot as president of Harvard (1869), and the foundation of Johns Hopkins University, a new kind of university focused on research and graduate studies under its inspirational first president, Daniel Coit Gilman (1876). Significantly, Owens College was one of the institutions visited by Gilman in September 1875 when he was drawing up his plans for Johns Hopkins.[4] No one would have thought of visiting ten years before.

The Mechanics' Institution – the 'Tech', as it became – underwent an extension of its own. The mechanics' institute movement of the first half of the century lost something of its rationale with the advent of free public libraries and universal elementary education, but Manchester's institute, under the visionary leadership of John Henry Reynolds (secretary, then principal, 1879–1912, and another Cross Street Unitarian), was able to take advantage of the recognition in the 1880s that Britain was lagging behind Germany and other countries in the provision of technical education. A Royal Commission on Technical Instruction sat between 1881 and 1884, with Owens College's Henry Roscoe as one of its members, investigating German technical education; at the same time, the Mechanics' Institution was reinvented as a Technical School (1883), and benefited from City and Guilds funding. When legislation in 1889 and 1890 empowered local authorities to fund technical education, the Technical School (after a brief period under the aegis of the Whitworth Institute) was taken over, along with all its assets, by the Manchester Corporation, and as the Municipal Technical School it moved to new premises on Sackville Street in 1902.[5]

The final part of this story of early foundations was the establishment of the Victoria University as a federal university in 1880 and of the independent Victoria University of Manchester in 1903, following which, in 1905, the Technical School acquired the status of the Faculty of Technology of the University. This was in itself a long and complex story. In brief: its self-confidence boosted in the wake of its 'extension', Owens College pressed from 1877 to be granted university status, to free it from the constraints of the curriculum of the University of London, for whose degrees many Owens students sat. Resistance from other northern cities, especially Leeds, led to the short-lived compromise of a federal university, eventually including both Yorkshire College Leeds and University College Liverpool. Once Birmingham University was founded in 1900 (hence its contested claim to be 'the first civic university', on the basis that a federation of civic colleges is not a civic university), campaigns began first in Liverpool and then in Manchester to break free from the federation.[6] Crucially, the arguments used reproduced those of the earlier campaign of the late 1870s: they rested on what the philosopher Samuel Alexander, in a noble manifesto for an independent university, called 'the growth of that sense of the organic connection of a university with its city and district'.[7] Subordination to the federal university, he argued, prevented the member colleges from forming a full and two-way relationship with their cities. Whether or not Manchester had the first of the civic universities, it was certainly the place where the idea of the civic university was born.

Plan of the book

Modern universities are, first and foremost, places of intellectual exchange, and yet university histories typically give short shrift to the life of ideas. There are some good reasons for that: intellectual histories are not readily confined within the boundaries of institutions, and university archives seldom give more than a fleeting glimpse of the clash of ideas. Still, historians have recently become far more attentive to *place*, and far more interested in institutions as sites where intellectual practices are shaped. So perhaps the time is right for an experiment in a new kind of university history, one which has as its focus the life of ideas in a university and the relationship of those ideas to the place (city as well as university) within which they are formed.

The aim in planning this volume has not been to select the eighteen or twenty most significant thinkers or bodies of ideas with a connection to the University or its predecessors, still less those most worthy of celebration. We have many of the big names (Roscoe, Alexander, Namier, Polanyi, Blackett, Turing, even Cox) but not all (no Jevons and no Rutherford, for instance). We have tried to balance some iconic names with others who have been neglected or whose connections with Manchester have been forgotten. The rationale for the choice is to produce a diverse collection of studies that individually and collectively tell us something significant about intellectual life and intellectual innovation within the academy, and what they owe to their interface with their civic environment. In some cases the connection with place springs from intellectual encounters within the University (Chapters 9, 11 and 13, for instance); in others, it takes the form of engagement with the needs of the city (Chapters 6 and 12, and the vignette on Catherine Chisholm). Several chapters consider the transmission of ideas across national boundaries (Chapters 2 and 14, and the third and fourth vignettes), and the limits on such transmission. The contributors have been invited in most cases to think not principally about disciplinary histories but about their protagonists as public intellectuals or as civic activists. Thus David Hayton (Chapter 8) examines Manchester and Zionism through the public careers of the philosopher Samuel Alexander, the historian Lewis Namier and the chemist and first president of Israel Chaim Weizmann, while Mary Jo Nye (Chapter 9) explores the entangled lives of Patrick Blackett and Michael Polanyi: both of them eminent scientists and public intellectuals, close friends but with radically different political stances.

This design has some consequences. Inevitably a book centred on the intellectual life of universities is mostly about academics. This is not because students do not form part of and help shape that intellectual life, but because the intellectual life of the student community is evanescent and hard to recapture from the archive. In Chapter 16, however, Joe Earle, Sukhdev Johal and Karel Williams make an intriguing attempt to write a kind of contemporary history of ideas focused on students, one of the authors (Earle) having been himself a leading participant in the student movement to reshape the teaching of economics. The authors argue trenchantly that the 'Post-Crash' movement was not about left-wing or heterodox economics, but simply about pluralism in the discipline. The movement was making the case for liberal education; and even if it was only partially successful in changing the economics curriculum, for many of the participants the experience of campaigning for change was a liberal education in itself. There remain questions to be explored about why economics as a discipline responded in a markedly aberrant way to the pressures of research assessment and university managerialism.

Several contributions focus on figures on the edge of the University community. In Chapter 4, Elizabeth Gow portrays the philanthropist and library-creator Enriqueta Rylands as a creative figure who had a powerful vision for her library, its connections with the University, and the relationships of both with the wider civic community. John McAuliffe reconstructs the fascinating story of the work of Esther Roper (an Owens College graduate and active suffragist) and the Anglo-Irish poet Eva Gore-Booth, as Associates of (and voluntary workers at) the Manchester University Settlement, a Ruskinian initiative founded on the model of Toynbee Hall in the East End. Among much else, McAuliffe picks up on Roper's insight that Gore-Booth's 'most characteristically Irish poems were written in Manchester': the opposition between the natural world of Sligo and the rampant modernity of the industrial city was an important theme. Roper belonged to the early cohort of women graduates from Owens College. Catherine Chisholm came just a few years later, but was notable as the first woman to graduate in medicine. As Peter Mohr's work shows, her subsequent career in child health was innovative in numerous ways. The Manchester Babies' Hospital,

7

her greatest achievement, was the product of her close engagement with an important group of enterprising women who were active in Manchester's public life in the early twentieth century, including Margaret Ashton and Shena Simon.

It is, of course, no accident that in the first half of the volume women appear only on the fringes of the academy. Women were admitted to the Mechanics' Institution from 1837. They were, however, excluded from Owens College by the terms of John Owens's will, and although that restriction was removed by the Owens College Act of 1871, the college did not admit women until 1883. Even a century later the number of women in senior academic positions was very small. Dorothy Emmet, the subject of Chapter 13, was not quite the first woman professor at Manchester, but throughout her twenty years as Professor of Philosophy she was the sole woman professor in the University, and indeed the sole woman on Senate. UMIST did not have a woman professor until 1995.[8]

Contests over the idea of the university form an important theme in the book, and several of the chapters are directly concerned with moments when the relationship between universities and their stakeholders was being rethought. In Chapter 2, Peter Morris and Peter Reed consider the chemist Henry Roscoe's key role in transmitting German ideas of the university into English academic practice. It was Roscoe, the leading proponent of the reform or 'extension' of Owens College, who persuaded the trustees to send him and the principal, Joseph Greenwood, on a tour of German and Swiss universities in 1868, and their report informed the redesign of the college in important ways. Manchester's large and influential German community made the city receptive to these ideas. Chapter 3 then goes on to draw out the significance of the new kind of institution forged by the Owens College reformers of 1870: two of the protagonists, James Bryce and Thomas Ashton, had (like Roscoe) studied at Heidelberg.

Chapters 13 and 14, meanwhile, rediscover the contrasting academic vocations of two unduly neglected figures in the post-1945 University. Dorothy Emmet is a particularly intriguing figure in a book of this kind. Intellectually her work was profoundly shaped by interdisciplinary exchange at Manchester, especially with the social anthropologist Max Gluckman, the political scientist Bill MacKenzie, the economist Ely Devons and the chemist-philosopher Michael Polanyi; and she said that she wrote *Rules, Roles and Relations* (1966) 'because I believe that moral philosophers and sociologists can have things of mutual relevance to say to each other'.[9] But Rachael Wiseman shows that her conception of human beings both as personas (role-players) and as persons (creative individuals) also drew insight from her own performance of the role of dean of the Arts Faculty (1961–63), and if this experience did indeed help her to produce *Rules, Roles and Relations* it was an unusually fertile deanship. Emmet was deeply immersed in the life of Manchester; Gilbert Gadoffre, though clearly drawn to the University (he spent three distinct periods in the French department, in 1938–40, 1954–63 and 1966–78), conceived his role as akin to that of a cultural ambassador, as Derek Robbins's chapter explains. The *Guardian*'s obituarist wrote that Gadoffre tried to inculcate in British undergraduates 'the art of dialectics, axiomatic for French students but totally alien to the British'.[10] A veteran of the French Resistance, he had some experience of apparently forlorn causes. He may be the only Manchester professor whose chair application recorded that he had been condemned to death *in absentia* by the German military authorities.

Both Luke Georghiou in the final chapter and Nancy Rothwell in her epilogue place civic engagement at the heart of the contemporary University's distinctive commitment to social responsibility. But what do the contributions to this volume have to tell us about the extent of engagement with civic life over the past two hundred years, and the different forms it has taken? In the early period – prior to the establishment of Owens College in 1851 – it would be a category error even to try to think about civic engagement. There was no academic profession or academic community in Manchester, and both the Mechanics' Institution and the Medical School were outgrowths of a wider civic culture. Following the creation of Owens College an academic community was formed, and it became much more visible and self-aware after the reformed college relocated to

Oxford Road in 1873.[11] It was also committed to being outward-facing. Henry Roscoe was prominent in the cause of public engagement with science: an active member of the British Association for the Advancement of Science, he presided when it met in Manchester in 1887. It was Roscoe who conceived and organised eleven series of 'Science Lectures for the People' in Manchester between 1866 and 1879: in the first instance these were delivered by Owens College professors, but later Roscoe drew in national luminaries such as T. H. Huxley and Alfred Russel Wallace.[12]

One of Roscoe's best-known lecturers was William Boyd Dawkins, Professor of Geology and a central figure in the development of the Manchester Museum. He spoke 'On Coal' at the Hulme Town Hall in November 1870, and again on 'Our earliest ancestors in Britain' in the Public Hall in Collyhurst in January 1879. As Chris Manias explains, Dawkins is one of those Victorian notables whose work will often jar with readers today, not least because he framed his account of human prehistory in highly racialised terms. But he is also a singularly interesting case study in both the impact of scientific research and in public engagement with science in the late Victorian period. He was advisor to an early attempt to bore a Channel Tunnel, which though unsuccessful led to the discovery of the Kent coalfield; he was consultant to the Humber Tunnel project, and advised on water supply in several towns. Later he advised diamond mining firms in South Africa, oil shale prospectors in Australia, and marble quarries in Italy. He was especially interested in museums as vehicles for public engagement with scientific research. Dawkins distinguished his 'New Museum Idea', realised in the Manchester Museum, from the old notion of displaying curiosities to 'excite wonder, horror or disgust'.

Systematic connections between academic research and the city's government or economy were rare in the Victorian period and remained so well into the twentieth century. Michael Worboys demonstrates in Chapter 6, however, that Sheridan Delépine, the Swiss-born pathologist and bacteriologist, was strikingly innovative in this regard. During his thirty-year tenure of the new chair of Pathology and Morbid Anatomy he built up a public health laboratory

which was astonishingly active in providing diagnostic and investigative services not just in Manchester but for local sanitary authorities
across the north of England. He had, wrote the *Manchester Guardian*,
'rendered the most distinguished service in all that relates to the
hygiene and well-being of the community in and around Manchester';
but he died at a time when the financial basis of the services his
laboratory provided to local authorities was being questioned by
the University, which failed to exploit his legacy to the full.

Over the last two hundred years Manchester has been a place
where great wealth has existed in close proximity to extreme
poverty.[13] The interface between the academy and the social problems
of the industrial city is explored from different points of view in
Chapter 6, on cooperation between University and the city corporation in public health; in Chapter 7, located in the University Settlement in Ancoats; and in the vignette on Catherine Chisholm's
work in the field of child health, where both the city council and
voluntary action loomed large. Manchester has also long had a
diverse population with substantial migrant communities: in the
nineteenth and early twentieth centuries German, Italian and
Armenian among others, and the largest Jewish community in Britain
outside London. In the postwar period that diversity became more
marked with the arrival of large immigrant populations from south
Asia and the Caribbean. In Chapter 10, Gerardo Serra studies the
trajectory of Britain's first Black professor, Arthur Lewis, who worked
closely with the Manchester Afro-Caribbean community during his
decade-long tenure of the Jevons Chair of Political Economy. His
fundamental work on development economics, for which he was
awarded the Nobel Prize in Economics in 1979, was crucially shaped
not just by the University as an intellectual community, but also
by his experience as a social activist in Manchester. Lewis was a
significant figure in the Pan-African movement, but he seems not
to have attended the famous fifth Congess, held in Manchester in
1945, when he was still at the LSE.

In the postwar period the social sciences grew very rapidly at
Manchester and became one of the University's notable strengths.
This was in line with national government policy: the Clapham

Committee, which reported in 1946, recommended a major expansion of provision for social and economic research in the universities. Katherine Ambler's chapter on the Manchester School of Social Anthropology offers a new perspective on this process of discipline creation and its connection with the social context. The Manchester School is usually depicted as centred on Africa and 'under-development', and this was indeed a principal focus. But Max Gluckman was appointed to head a Department of Social Anthropology *and* Sociology, and Ambler shows that the intention was that this should be centrally concerned with the study of modernity, and capable of making a practical contribution to the challenges of industrial and urban modernity in Britain. She thus helps illuminate the 'Manchesterness' of the Manchester School: it was shaped by a university with a strong commitment to social science with a practical edge, and by its location in a city which for more than a century had been a byword for industrial modernity.

In the public mind, however, at this time the University's name was chiefly associated with big science such as the development of the Jodrell Bank Observatory and the construction of the world's first stored-program computer. Alan Turing, now a figure of iconic significance in the history of computing and much else, was also at the University in this period, as Reader in Mathematics (1948–54), but he arrived shortly after the construction of 'the Baby', with which he had no involvement. James Sumner's essay addresses the formidable task of weaving together a lucid account of what was Mancunian in the development of computing, and an account that makes sense both of the engineering tradition associated with Freddie Williams and Tom Kilburn and the mathematical-philosophical tradition identified with Turing.

If, in the public mind, Manchester science in the 1950s meant the computer and Jodrell Bank, in the 2020s it means graphene and Brian Cox. Luke Georghiou shows how central the graphene Nobel Prize – awarded in 2010 for a decisive breakthrough in 2004, the year of the merger – was to the self-image of the formally new University that was created by that merger. It also gave rise to a new kind of collaboration with civic leaders and national government to ensure the maximum economic exploitation of the isolation of graphene. Matthew Cobb's chapter, in contrast, explores the new kind of science communication practised by Cox. This is public engagement for the age of celebrity: not, in truth, engaging much with the city, but making prolific use of Jodrell Bank, an iconic venue firmly identified with the University of Manchester in the public consciousness since the 1950s. What emerges powerfully from Cobb's account is the importance of the conjuncture of September 2008, when the Large Hadron Collider (LHC) was switched on and propelled the rock star physicist Cox into a new career as media science communicator. Curiously, this all happened as Lehman Brothers teetered on the brink of bankruptcy, and one can only speculate what might have happened if it had gone over the edge on the LHC's big day.

The importance of Jodrell Bank in this story is worth highlighting: from an early stage in its history it had been recognised for its potential to open up radically new opportunities for engaging the public with scientific research in universities. As Shena Simon (an influential figure both in University and civic life) told the vice-chancellor in 1964, it offered the University the chance 'to do something rather dramatic which would impinge on the citizens of Manchester and make them proud of having the University at their doorstep'.[14] Bluedot – Jodrell's music and science festival, building on Cox's prowess at new kinds of science communication – may not be quite what she had in mind.

In all there are seventeen full chapters, complemented by four shorter vignettes that tell more focused stories. They are arranged chronologically, and divided into three parts. The dividing lines are at 1903–05 (the creation of the independent Victoria University of Manchester, and its assimilation of the Technical School as its Faculty of Technology) and 2004 (the creation of the new University of Manchester, following the merger). The three parts thus correspond approximately to the nineteenth, twentieth and twenty-first centuries. Each part begins with a brief introductory section that surveys the development of the University in connection with the development of the city. Together, these introductory sections trace the evolving sense of what it means to be a university *for* Manchester as well as *of* it.

⋆ ⋆ ⋆

I am grateful to Luke Georghiou and Charlotte Wildman for their comments on a draft of this introduction, and the other editorial matter.

Notes

1 G. W. Daniels, 'Economic and commercial studies in the Owens College and the University', *The Manchester School*, 1 (1930), 3–4.
2 'Town', because Manchester acquired city status only in 1853. When the Mechanics' Institution was founded in 1824, Manchester was not yet an incorporated borough, a status it acquired, after a struggle, in 1838.
3 For Heywood, see 'The County Election, departure of Mr Heywood for Lancaster', *Manchester Guardian*, 14 May 1831, p. 3; for Greg, see 'Sir Thomas Hesketh's canvass', *Manchester Courier*, 22 December 1832, p. 4.
4 Fabian Franklin, *The Life of Daniel Coit Gilman* (New York: Dodd, Mead & Co., 1910), p. 209; *Addresses at the Inauguration of Daniel Coit Gilman, as president of the Johns Hopkins University, Baltimore, February 22, 1876* (Baltimore, MD: Murphy, 1876), p. 52.
5 On its transfer to the Corporation the Technical School had an endowment income of £4 per year: 'Manchester Corporation and the Technical School: important recommendations', *Manchester Guardian*, 22 January 1892, p. 8. In addition, the value of the property the Corporation acquired from the Technical School and the School of Art together contributed £31,000 towards the cost (estimated at £125,000) of the Sackville Street building eventually opened in 1902: Henry Roscoe, 'The Manchester Municipal Technical School', *Nature*, 47 (1892), 203.
6 Eric Ives, Diane Drummond and Leonard Schwarz, *The First Civic University: Birmingham 1880–1980* (Birmingham: Birmingham University Press, 2000).
7 Samuel Alexander, 'A plea for an independent university in Manchester', *Manchester Guardian*, 11 June 1902, p. 12.
8 I am grateful to the labour economist Jill Rubery for confirming that she was indeed the first at UMIST. The first at the Victoria University was Mildred Pope, Professor of French Language and Romance Philology, 1934–39.
9 Dorothy Emmet, *Rules, Roles and Relations* (London: Macmillan, 1966), p. ix.
10 Geoffrey Harris, 'Cultural ambassador', *Guardian*, 30 March 1995, p. A15.
11 Colin Lees and Alex Robertson, 'Characteristics of professional life in the first of the civic universities, 1851–1918', *BJRL*, 82:1 (2000), 225–50.

12 David Riley, 'The Manchester Science Lectures for the People, c. 1866–79', *BJRL*, 85:1 (2003), 127–45. 'Science' was conceived broadly, since the lectures included the economist Stanley Jevons on coal and Adolphus Ward on Charles Dickens.

13 There were, however, very few Mancunian millionaires in the nineteenth century: the greatest fortunes were London-based; see W. D. Rubinstein, 'The Victorian middle classes: wealth, occupation, and geography', *Economic History Review*, n.s. 30:4 (1977), 602–23, and many other works by Rubinstein.

14 Manchester Central Library, Shena Simon papers M14/2/3/8, Lady Simon to Sir William Mansfield Cooper, 9 June 1964.

PART I

Academy and community in the nineteenth-century city

Stuart Jones

It was in the nineteenth century that Manchester established itself as one of the world's great cities, when its name became synonymous with industrial capitalism, in both its positives and its negatives. In a striking essay, Janet Wolff has called Manchester (not Paris, as Walter Benjamin argued) 'the capital of the nineteenth century'.[1] It was the city that across the world came to epitomise the new kind of civilisation brought into being by the Industrial Revolution. There was even a German word for it: *Manchestert(h)um*, generally used pejoratively by those who thought this new kind of civilisation was no kind of civilisation at all. It meant free trade, global connectivity and huge fortunes, but also unprecedented population growth, overcrowded slums and air pollution: 'Manchester devil's darkness', as John Ruskin called it.[2]

The development of higher education institutions was part of an attempt by the burghers of Cottonopolis to show that there could be high culture even in a smoke-filled industrial city founded on the values of free trade, and from the 1860s onwards the mercantile elite, who had previously been sceptical of the uses of the higher learning, now saw that a city of Manchester's importance must have a university. Of the lay figures in the Owens College extension (Chapter 3), Thomas Ashton had played a leading role on the executive committee for Manchester's great Art Treasures Exhibition of 1857, and Robert Darbishire and Alfred Neild were local secretaries (along with Roscoe) when the British Association for the Advancement of Science met in Manchester in 1861.[3] These were important opportunities to put Manchester on the cultural and scientific map, and the extension of Owens College was a much more ambitious project with a similar aim. It was the product of a new wave of assertive civic pride that also bore fruit in the new Town Hall, built 1868–77, and the new Royal Exchange, opened in 1874.

Religion was fundamental to social identities in Victorian Britain, and inter-denominational rivalries dominated educational developments. Manchester was a city of religious diversity, with substantial Roman Catholic, Protestant Nonconformist and Jewish communities, alongside an Anglican elite that retained its institutional strengths. There was a strongly Liberal political consciousness among the Protestant Dissenters, who were numerous in the business community. The key role played by the Unitarians of Cross Street Chapel in founding the Mechanics' Institution and other cultural institutions of the time has already been noted. They were also prominent as trustees of Owens College and, later, as leaders of the extension movement. An important driver in the movement for higher education in Manchester was the existence of a wealthy Nonconformist mercantile elite whose sons were formally or informally excluded from Oxford and Cambridge on religious grounds. Not until the Universities Tests Act of 1871 were the two old universities fully opened to non-Anglicans on equal terms. Owens College was determinedly non-denominational from the outset, and indeed excluded the teaching of theology until the creation of the independent Victoria University of Manchester in 1903. If Owens College had models for the kind of education it offered, they were drawn from University College London, from the Scottish universities, and later from Germany.

When the founding president of Johns Hopkins University visited Owens College in 1875, he was struck by the excellence of its laboratories. The college developed an early strength in those disciplines – chemistry and engineering in particular – which were most closely aligned with the needs of local industry. That was one aspect of what the emergent idea of the civic university implied, certainly in comparison to Oxford and Cambridge, where science was marginal before the 1860s. The distinction of Owens College scientists is incontestable: of twelve professors in 1872–73 (those shown in Figure 3.2 on p. 65), four were current Fellows of the Royal Society, and one other was elected in 1877; and there was another FRS, Schorlemmer, as a rather over-qualified laboratory assistant (he was made a professor in 1874). By 1880, when the Victoria University was founded, Owens had eight Fellows of the Royal Society out of twenty-four professors.

But it is wrong to imagine that what made the English civic universities distinctive was a preponderant emphasis upon science and engineering. Peter Burke, in his admirable *Social History of Knowledge*, writes that Owens College was 'originally founded by a local textile merchant to teach practical subjects'.[4] This was true enough of some later civics such as Sheffield, but not at all true of Manchester or indeed Liverpool.[5] John Owens was quite explicit about his intended subject coverage: the college should instruct 'young persons of the male sex … in such branches of learning and science as are now and may be hereafter usually taught in the English Universities'.[6] His trustees were punctilious in following this injunction, establishing chairs in classics, history and philosophy from the outset. It is well known that the chemist Henry Roscoe did more than anyone to shape the intellectual reputation of the college, and also played a decisive role, along with scientifically trained industrialists such as Thomas Ashton, in the movement for the extension of Owens College in 1870.[7] But the four principals of Owens were all humanities scholars, and the Senate in the wake of the college extension was equally balanced between the sciences and the humanities.

This breadth mattered to lay opinion as well as to the academics. Richard Copley Christie, first a professor and then a lay member of Council, was notably important in shaping Owens College as a place of humane learning, He funded the construction of a library, named the Christie Library, and left his own collection which made the library a major resource for research in Renaissance studies.[8] The largest single benefaction to higher learning in Victorian Manchester was not John Owens's bequest, nor Charles Beyer's endowment of science professorships, but Enriqueta Rylands's establishment of the John Rylands Library. Though the Rylands was endowed as an independent library open to the general public, and remained independent until it merged with the university library in 1972, University professors were closely involved in its management from the outset. Conversely, Mrs Rylands was a central figure in the establishment of the Faculty of Theology in 1904: an achievement of real significance, since as a truly non-denominational faculty it was acclaimed (with a hint of exaggeration) as 'perhaps the first faculty of theology in the world which can be called a real university faculty'.[9]

Notes

1 Janet Wolff, 'Manchester, capital of the nineteenth century', *Journal of Classical Sociology*, 13:1 (2013), 69–86.
2 John Ruskin, *The Storm Cloud of the Nineteenth Century* (Orpington: George Allen, 1884), p. 56.
3 'British Association for the Advancement of Science', *Observer*, 9 September 1861, p. 3.
4 Peter Burke, *A Social History of Knowledge, vol. 2: From the Encyclopédie to Wikipedia* (Cambridge: Polity, 2012), p. 134.
5 On this contrast, see Michael Sanderson, *The Universities and British Industry 1850–1970* (London: Routledge and Kegan Paul, 1972), ch. 3. But Sanderson tends to overstate the quest for direct economic benefit as a motive for the creation and expansion of the civic universities, and to understate civic pride.
6 Quoted by William Whyte, *Redbrick: A Social and Architectural History of Britain's Civic Universities* (Oxford: Oxford University Press, 2015), p. 93.
7 W. H. Chaloner, *The Movement for the Extension of Owens College Manchester 1863–73* (Manchester: Manchester University Press, 1973); also Henry Enfield Roscoe, *The Life and Experiences of Sir Henry Enfield Roscoe* (London: Macmillan, 1906), pp. 110–11.
8 'The Christie Library and Whitworth Hall, Owens College', *Manchester Guardian*, 21 June 1898, p. 12; and Moses Tyson, *The Manchester University Library* (Manchester: Manchester University Press, 1937), p. 16.
9 H. B. Charlton, 'University education in Manchester', *Manchester Guardian*, 16 May 1938, pp. A20–1.

The mechanics' institutes and the spread of 'useful knowledge'

Joanna Bourke

On 18 October 1858, statesman, orator and poet George Howard (the 7th Earl of Carlisle) stood before members of the Manchester Mechanics' Institution to distribute the annual certificates of achievement to the students. His speech proved popular with those present, who loudly cheered and clapped when Howard commended them for their educative labours. It was 'a most pleasing thing', he contended, that a town such as Manchester, which was 'foremost ... in all the world for mechanical invention and ingenuity' and therefore 'absorbed necessarily to a great extent in material pursuits', should also have made 'suitable provision ... for the intellectual, the mental, and the higher qualities of our nature'. In the thirty-four years since its foundation, he marvelled, the Institution had flourished. From a small number of male students in 1824, it now boasted 441 day-time students (half of whom were women), and over a thousand male students attending in the evenings. The success of the Institution reminded Howard of the limitations of his own education, decades earlier. In particular, he lamented not having had 'the advantage of being taught by means of the chalk and black board – (laughter and cheers) – which he learned was such an efficient instrument'. After all, blackboards had only been invented in 1801 by Edinburgh headmaster James Pillans. Howard urged the students to use the knowledge that they gained at the Mechanics' Institution to 'attain to the highest degree of usefulness to the generation in which you live'. After the speech, Howard distributed ten 'certificates of merit' to the students, whose occupations were listed as a packer, a wood-turner, a warehouseman, two bookkeepers and five engineers.[1]

Howard's speech draws attention to some of the features that animated the establishment of mechanics' institutes throughout the UK and the world. These include the dramatic growth in educational establishments, the importance of 'useful knowledge', and a belief in the role of education in developing not only people's intellectual qualities but their moral and spiritual ones as well. The Manchester Mechanics' Institution had been established in 1824, a time of rapid growth in the number and popularity of mechanics' institutes. Although classes for the education of working people had existed sporadically since the eighteenth century (most notably, Joseph Middleton's Spitalfields Mathematics Society, which taught scientific subjects to artisans and craftsmen from 1717), the *systematic* establishment of mechanics' institutes is often credited to physician and philanthropist George Birkbeck (plate 2). In 1799 he had been invited to teach 'natural philosophy and chemistry' at the Andersonian Institution, which had been founded three years earlier under the will of John Anderson, former Professor of Natural Philosophy at the University of Glasgow. Needing a particular machine for his classes, Birkbeck had visited a mechanical workshop and was struck, first, by the workmen's ignorance of basic engineering facts and, second, by their hunger for knowledge. He promptly opened his classes to mechanics, offering classes on Saturday evenings. Birkbeck's aim was to ensure that a working man would 'cease to be a mere machine, toiling on from day to day' but would 'understand the laws on which his operations were based' and therefore 'perfect himself in his calling'.[2] Demand was high; by the fourth class, there were five hundred men in attendance.

George Birkbeck left Glasgow for London in 1804, where he set up a fashionable medical practice, in addition to working as a physician to the General Dispensary for the Relief of the Poor in Aldersgate Street, which provided home-visiting and outpatient treatment for paupers and working-class men and women in that area of London.

Birkbeck was also active in the Medical and Chirurgical Society, the Chemical Society, the Meteorological Society and the London Institution, established in 1809 for the diffusion of science, literature and the arts. It was during this stage in his life that he forged friendships with radicals and economists including David Ricardo, Joseph Hume, William Cobbett, Jeremy Bentham, George Grote, James Mill and John Stuart Mill. Within this milieu, George Birkbeck was well placed to throw his energies into a wide variety of social causes, including support for the Reform Act of 1832, which transformed the electoral system, and the fight to repeal the duty on paper and the tax on newspapers, which severely limited the ability of people to contribute to political debate and disseminate knowledge. In 1823, when patent agent Joseph Clinton Robertson and economist Thomas Hodgskin (both of the *Mechanics' Magazine*), along with the 'radical tailor' Francis Place, put out a call for the establishment of a London Mechanics' Institution, Birkbeck was considered the ideal president. It was the start of what would become an international movement. By the middle of the nineteenth century, when Howard was giving his oration to the members of the Manchester Mechanics' Institution, there were an estimated 610 mechanics' institutes in England, 55 in Scotland, 25 in Ireland and 12 in Wales.[3] Institutes were also flourishing in the US, Canada, South Africa, Australia, New Zealand and elsewhere.

These institutes claimed to address the educational needs of 'mechanics', but this term was used much more widely than it is today. Although most mechanics' institutes were established with the support of local dignitaries and philanthropists, the majority of 'members' (they were called 'students' much later) called themselves 'operatives' or 'members of the working class'. They were not 'labouring poor'. There were heated debates about who should wield power within the institutes. In one article, published in 1826 and entitled 'To the Members and Managers of the Mechanics' Institutions in Britain and Ireland', the anonymous author insisted that the 'operative classes ... *for whose exclusive benefit all Mechanics' Institutions profess to be instituted*' (emphasis in the original) must retain control. The author admitted that

The employers of the operative classes, as well as others of the benevolent rich, have in many places come forward with their money, books, and instruments, to aid in the establishment of your institutions. Such is one of the best modes of promoting mutual kindness between the rich and the productive classes. But let not this class of person expect *power* in return for their gifts. By so doing they would have made sordid bargains instead of gifts, and would nullify all those claims to beneficence and sympathy, which would be otherwise their natural and sufficient reward.[4]

However, as the century progressed, the proportion of members who 'worked with their hands' declined: the classes were increasingly populated by clerks, shopkeepers and teachers, for example. This shift in membership led to a broadening of the curriculum. Technical and scientific subjects remained important, but more literary, historical and arts-based classes also proved popular. As Howard quipped in his address to the Manchester Mechanics' Institution, while mechanical, engineering and other scientific pursuits were being taught, they were being supplemented with classes in 'music, dancing, drawing, modelling, French and even millinery – no doubt an important pursuit in its way – (laughter)'. Howard's banter and the laughter that followed was a wry reflection of the men's unease at the educational needs of the female members of the Institution.

The involvement of women in the mechanics' movement was a source of anxiety in the early decades. On the one hand, would the presence of women in the institutes lead to immorality, as un-chaperoned women mingled with men in the same halls? Would it enhance competition for books and lecture space? Would the 'scientific' quality of the lectures be diluted, since it was feared that women might demand more leisure-oriented classes such as art and music? On the other hand, would educating working women enhance their employment opportunities (for example, as teachers of art and music)? Would it promote more fulfilling and equitable relationships between men and women? As early as 1826, prominent Owenite William Thompson pleaded with the mechanics' institutes to

Let your libraries, your models, and your lectures … be equally open to both sexes. Equal justice demands it… Long have the rich excluded the poorer classes from knowledge; will the poor classes now exercise the same odious power to gratify the same anti-social propensity – the love of domination over the physically weaker half of their race?[5]

This was a point echoed by the radical politician Rowland Detrosier in 1831. Detrosier had founded the New Mechanics' Institution, a breakaway from the Manchester Mechanics' Institution, and he believed that the education of women had to be central to working-class emancipation. He argued that it was necessary to 'raise the

1.1 A woman running up the steps of Breams Building, Chancery Lane, the home of the Birkbeck Literary and Scientific Society. In many mechanics' institutes there was considerable anxiety about women attending classes. Courtesy of Birkbeck, University of London.

females of the working classes from the state of degradation' in order to make them 'rational companions of men'. By providing working-class women with education, their menfolk would be encouraged to turn away from 'immorality and crime'.[6] Educated women would raise rational children; their domestic labour would enhance families, communities and the entire nation. In Detrosier's words, 'individual reform will secure national happiness'.[7]

For liberal as well as radical educationalists, the lodestone was 'useful knowledge'. No one promoted this theme more effectively than Henry Brougham. Brougham was a lawyer by profession, but he came to public notice through his role as chief advisor to Queen Caroline, the estranged wife of King George IV. By the 1830s Brougham was Lord Chancellor of Great Britain, where he was an active supporter of the 1832 Reform Act and, as an active abolitionist, the 1833 Slavery Abolition Act. In 1826 he founded the Society for the Diffusion of Useful Knowledge. Crucially, however, his book on education was a popular success. Entitled *Practical Observations Upon the Education of the People. Addressed to the Working Classes and their Employers* (1825), it became the 'Bible' for the mechanics' institutes everywhere. In its first year alone, *Practical Observations* went through nineteen editions.[8] According to one admirer, not since 'the Scriptures were first printed and circulated in the common tongue' had there been such an important text.[9] In *Practical Observations*, Brougham contended that only 'tyrants' and other 'bad rulers' should be terrified by 'the progress of knowledge among the mass of mankind'.[10] He maintained that 'the time is past and gone when bigots could persuade mankind that the lights of philosophy' were 'dangerous to religion'. He argued that the 'peace of the country, and the stability of the government, could not be more effectually secured than by the universal diffusion' of knowledge about the 'true principles and mutual relations of population and wages'.[11]

Brougham's 1825 book was a milestone, in large part because the mechanics' institutes were facing exceptionally high levels of hostility at the time. Their opponents had four major concerns. The first was that educating working people would be detrimental to good morals. John Dunlop, a Scottish Justice of the Peace, maintained

PRACTICAL OBSERVATIONS

UPON THE

EDUCATION OF THE PEOPLE,

ADDRESSED TO

THE WORKING CLASSES

AND

THEIR EMPLOYERS.

BY

H. BROUGHAM, Esq. M.P. F.R.S.

LONDON:

PRINTED BY RICHARD TAYLOR, SHOE-LANE;

AND SOLD BY LONGMAN, HURST, REES, ORME, BROWN, AND GREEN,
PATERNOSTER-ROW,

FOR THE BENEFIT OF THE LONDON MECHANICS INSTITUTION.

1825.

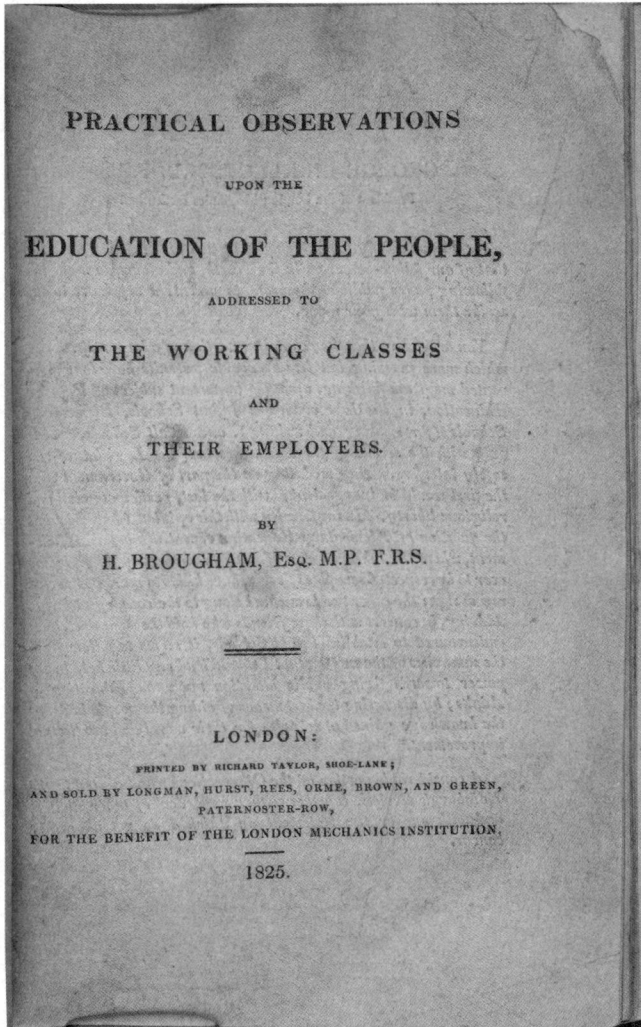

1.2 Henry Brougham's 1825 book *Practical Observations Upon the Education of the People* was considered the 'Bible' of the mechanics' institute movement. Courtesy of the John Rylands Library.

that a 'literary and scientific education' would not 'generate that change of heart required in Scripture', nor would it 'advance morality to a sublime and scriptural pitch'.[12] He was especially worried about how education could be 'beneficial to morality' for female members.[13] Scottish essayist and episcopalian priest Archibald Alison expressed similar sentiments in more vivid terms. In 1838 he argued that educating working men would give them 'the means of the gratification of the animal or the sensual propensity'.[14] He also advocated for censorship of the books made available to working people. He contended that education that was 'unaccompanied ... by any adequate restriction upon the books [workers] read or any adequate religious instruction' was a 'very great cause of the depravity of the times'.[15] Alison agreed with philosopher Francis Bacon that 'knowledge or education is power', but it could be the power to do mischief as well as good.[16] In his book *The Social Condition and Education of the People in England and Europe* (1850), Joseph Kay summarised this position when he observed that

> a spirit omnipotent for evil, a spirit of revolution, irreverence, irreligious, and recklessness, and, more dangerous of them all, a spirit of unchecked, unguided, and licentious intelligence is abroad, which will be the most dangerous enemy, with which Christianity had hitherto had to cope.[17]

The second, related concern expressed by opponents of the mechanics' institutes was that they were a threat to the Church of England's monopoly over education. Churchmen and their supporters feared that any education that was primarily 'scientific and philosophical' rather than 'moral and religious' was risky, as conservative Anglican biblical scholar Edward William Grinfield warned in 1825.[18] Grinfield believed that 'it would be far better that the common people of this country should remain totally illiterate, than they should thus be furnished with *tools* by which they would inevitably work out their own and the public ruin'.[19]

Like Alison, Grinfield wanted to know *who* was going to be responsible for choosing what books working men would be reading in mechanics' institutes. He believed that it was 'desirable that the choice of such books should be left to those whose superior knowledge may enable them to direct the reading of others'.[20] It was 'folly' to educate people beyond what they needed to know, especially when such knowledge would not make a working man 'more happy in that station to which Providence had called him'.[21] Grinfield urged the 'upper classes of society' to do every thing in your power to give the labouring orders a religious, virtuous, and *useful* education, by founding this education on the love and the fear of God, and by associating it with a strong attachment to the *existing* institutions of our country.[22]

This sense of 'useful' was very different from the one promoted by people such as Brougham and Birkbeck. For Grinfield, 'useful' meant ensuring the maintenance of social hierarchies. He addressed working people directly, advising them to

remember that knowledge is valuable only as it is connected with an advancement in piety and religion; that sobriety and contentment are far more valuable qualifications than any attainments in mere art or science; and that those who would confine your education *exclusively* to the purposes of the present life are no real friends to your happiness and welfare.[23]

Theological concerns were augmented by more practical ones. The third and fourth anxieties related to the establishment of mechanics' institutes was that they would foment social and political disorder. Some commentators alleged that there was a link between education and crime. A 'comparatively better education was co-existent with a greater amount of crime', claimed John Dunlop, speaking before the 1834 Select Committee on Drunkenness. If evidence were needed, he urged the commissioners to look at France.[24] Archibald Alison made a similar argument before the Select Committee on 'Combinations of Workmen' (that is, trade unions). He claimed that a statistical analysis of 83 departments in France revealed that 'the amount of crime is just in proportion to the quantity of intelligence that prevails; that crime is invariably the greatest in those departments where there is most knowledge and education, and invariably the least in the reverse'.[25]

The possibility that educated working people would incite social disruption, however, paled alongside the threat of political chaos. Of all the arguments against mechanics' institutes, this was the most impassioned. For Moses Angel, who was active in the Jews' Free School, the mechanics' institutes were an improvement on 'the tavern, the billiard-room, the cheap theatre, and the casino' because they at least *attempted* to encourage 'the cultivation of the mind'. However, they also risked turning 'half-educated men into ill-formed politicians (and therefore, generally revolutionists, or at least democrats)'.[26] For Angel, democracy was as dangerous to social stability as all-out revolution. This was also what Orthodox Chief Rabbi of the Empire Nathan Marcus Adler argued, although he framed the issue more broadly by musing that members of the institutes were turned into 'dreamers, talkers, and vague reasoners'.[27] One critic even contended that educating working men was as perilous as educating animals. 'Suppose', this anonymous author in the *Edinburgh Review* wrote in 1826, that 'some friend to humanity were to attempt to improve the condition of *the beasts* of the field; – to teach the horse his power, and the cow her value'. Wouldn't that make the animal less 'tractable and useful' and not 'so profuse of her treasures' (that is, milk) 'to a helpless child?'[28] Once again, 'useful knowledge' was equated with knowledge that maintained the status quo rather than questioning it. It was linked to anxieties about weakening the deference of the 'lower orders' towards their 'superiors'. In the words of an author writing in 1825 in the *St. James's Chronicle*, 'every step which they take in setting up the labourers as a separate and independent class' was a step towards destruction: 'A scheme more completely adapted for the destruction of this empire could not have been invented by the author of evil [that is, the devil] itself.' In *Mischiefs Exposed* (1826), the Revd George Wright, vicar of Askham Bryan, expressed this view succinctly, warning that the institutes were guaranteed to 'degenerate into Jacobin clubs, and become nurseries of disaffection'.[29]

This barrage of hostility towards mechanics' institutes and other educational organisations catering to working people failed to halt the momentum of reform. After all, the institutes were part of a much larger movement within politics at the time. The early nineteenth century was a time of intense political and economic reform. Civil society itself was undergoing revolutionary change. Industrialisation was transforming relationships between the different classes. Even the most conservative employers were beginning to recognise the need for more literate and educated workers if their businesses were to flourish. This was especially the case in large industrial centres such as Manchester where the economy was dependent on the skills of its workers. Economic competition, particularly from continental Europe, was a concern. Was the UK falling behind? Governments were tackling elementary education (the Education Act of 1870 was important), providing compulsory education and progressively raising the age to which children were required to attend school. The instruction given in mechanics' institutes was a useful supplement.

And working people themselves were keen. This was largely because the mechanics' institutes were very much *local* initiatives. Indeed, it may be problematic to write as though the rapid spread of mechanics' institutes throughout the UK and the wider world were part of a *movement*. It was more ad hoc than the term 'movement' implies. Mechanics' institutes were established by local communities; they served local interests. The institutes in Australia and Ireland can be taken as examples. Colonists in Australia quickly adapted what they knew about mechanics' institutes in Britain to the local context. Australian institutes were often based in under-developed, rural areas, such as Van Diemen's Land, which in 1827 established the first mechanics' institute in Australia. These institutes were much more than simply educational establishments. Their buildings served as libraries, reading rooms, meeting spaces, galleries and theatres, as well as venues for sports, fairs, weddings, baptisms and funerals. Similarly, communities in Ireland had begun forming mechanics' institutes from 1824. But again, the context in which they were established was very different to the context in industrialised cities such as Manchester or London. As in Australia, Irish institutes were located in provincial, agricultural towns. They played an important role in cutting across sectarian divides, providing relatively neutral spaces where Catholics and Protestants could mingle.[30] Many Irish institutes were inspired by temperance politics, as was the Rotherham Mechanics' Institute.[31]

Local needs varied. For example, the Ancoats Mechanics' Institute in Manchester focused very much on elementary reading and writing, in contrast to institutes in Newcastle, Nottingham and Sheffield, which provided higher-level scientific instruction.[32] Membership also varied. Most of the institutes in the north of England, for example, tended to attract more working-class members, but that in Wakefield emerged out of a debating club led by local middle-class savants.[33] This difference was commented on in the 1859 annual report of the Yorkshire Union of Mechanics' Institutes, which stated that

> It is a prevalent opinion that Mechanics' Institutes are only so in name, their original purpose having been superseded by the rejection of them by the class for whom they were intended, and their adoption by the middle classes. But this is not true of the majority of those in Yorkshire, however it might apply elsewhere. Some of the most flourishing Institutes are composed almost wholly of the labouring class, and in most of them they form a considerable majority.[34]

Some were under the leadership of paternalistic middle-class leaders, as was the case with the Bradford Mechanics' Institute;[35] others, such as the Yorkshire Union of Mechanics' Institutes, were dominated by the workers themselves.[36]

These differences were important because they reflected alternative visions of the purpose of mechanics' institutes. Were they vehicles for improving social morality or a key to promoting an entrepreneurial ideal? Was the spread of the institutes due to the enthusiasm of working people themselves to 'better' their position or was it more 'political' – that is, were they hotbeds for the promotion of universal suffrage, factory reform or even revolution? This diversity in aims has stimulated a lively debate about the value of the mechanics' institutes. Even vigorous *champions* of education for working people had concerns. Famously, in *The Condition of the Working Class in England* (1844), Friedrich Engels reproached the mechanics' institutes for being 'organs for the dissemination of the sciences useful to the bourgeoisie'. He accused their teachings of being 'tame, flabby, subservient to the ruling politics and religion'. Engels even denounced the institutions for propagating a 'constant sermon upon quiet obedience, passivity, and resignation to his fate'.[37] He believed that working people should be educated in proletarian reading rooms, run by the workers themselves, rather than those sponsored by wealthy men such as Birkbeck and Brougham.

This critique has been taken up by some historians. In the words of historian of science Steven Shapin, the men who established mechanics' institutes not only believed that 'a scientifically educated working class would aid the process of industrialization', but they also thought that a 'scientific education might prove to be an instrument of *social control*'.[38] This could involve subduing labour protests, encouraging workers to accept a particular version of economic management, or stifling sectarian conflict (as in Ireland).[39] The problem with the 'social control' argument is that it does not allow for the high levels of inconsistency in the actual practices of the men and women within the various mechanics' institutes, and it overstates the greater degree of conscious manipulation of working people by the managers. Some institutes – most notoriously the London Mechanics' Institution – expressively forbade any discussion of politics or religion.[40] However, this directive was not enforced in practice. In fact, the LMI rented out its rooms and lecture theatre to some of the most radical organisations of the time, including the Friends of Civil and Political Liberty, the Owenite London Co-operative Society, the Society for Promoting Radical Reform, and numerous trade unions, socialists and political radicals.[41] The mechanics' institutes in Leeds and Manchester promoted Chartist and Owenite ideas; in Cheltenham, the writings of Bentham, Owen and Cobbett were in the library; Sunderland welcomed Chartist leaders.[42] Furthermore, institutes that began with one motive – improving the 'morals' of working people – might develop into something much more emancipatory and socially diverse as the century continued.[43] Many of the institutes provided their members with vital organisational and administrative skills, which encouraged them to participate in political movements and radical causes – indeed, it gave them the necessary skills to do so effectively. As economic historian John Laurent argues, working people used education to 'transform agencies designed for social control into instruments for emancipation'.[44] They were a major propeller for the Labour movement, socialism, co-operation, social reform and enfranchisement.

The definition of 'useful knowledge' was a fundamentally contested one. Deeply embedded ideas about class differences – including fundamental beliefs about the innate superiority and inferiority of the different social 'orders' – meant not only that working people were viewed as incapable of imbibing knowledge but that, when they did so, it could only result in societal chaos. The founders of the various mechanics' institutes, as much as their opponents, had differing views about what was 'useful'. Their members or students also sought education for a variety of motives, including social advancement, political ambition and leisure. When addressing members of the Manchester Mechanics' Institution, George Howard (with whom I started this chapter) insisted that the Institution's aim was to develop 'the intellectual, the mental, and the higher qualities of our nature'. What that precisely meant was decided by working people themselves.

Notes

1 'Manchester Mechanics Institute', *Daily News*, 20 October 1858, p. 3.
2 John George Godard, *George Birkbeck. The Pioneer of Popular Education. A Memoir and a Review* (London: Bemrose and Sons, 1884), pp. 27–8.
3 James William Hudson, *The History of Adult Education, in Which is Comprised a Full and Complete History of the Mechanics' and Literary Institutions, Athenæums, Philosophical, Mental and Christian Improvement Societies, Literary Unions, Schools of Design, Etc., of Great Britain, Ireland, America, Etc. Etc.* (London: Longman, Brown, Green and Longmans, 1851), p. v.
4 'To the members and managers of the Mechanics' Institutions in Britain and Ireland', *Co-Operative Magazine and Monthly Herald*, 1:1 (January 1826), 23–4.
5 William Thompson, 'To the members and managers of the Mechanics Institutions in Britain and Ireland', *Co-Operative Magazine and Monthly Herald*, 1:2 (February 1826), 46.
6 Rowland Detrosier, quoted in 'London Mechanics' Institution', *The Examiner*, 25 September 1831, p. 619.
7 Ibid.
8 These sales figures are according to 'A reply to Mr Brougham's "Practical Observations Upon the Education of the People, Addressed to the Working Classes and Their Employers"', *Edinburgh Review*, 42:83 (1 April 1825), 212.
9 'Review of Brougham's tract entitled "Practical Observations Upon the Education of the People; Addressed to the Working Classes and Their Employers" (London, 1825)', *Edinburgh Review*, 41:82 (1 January 1825), 508.

10 Henry Brougham, *Practical Observations Upon the Education of the People. Addressed to the Working Classes and their Employers* (1825) (Manchester: E. J. Morten, 1971), p. 31.

11 Ibid., p. 4.

12 Evidence from John Dunlop in the Report from the Select Committee on Inquiry into Drunkenness, with Minutes of Evidence and Appendix (5 August 1834), PP 1834 [559], p. 412.

13 Evidence from Revd Dr Nathan Marcus Adler, in Popular Education Commission, volume 5, Answers to the Curriculum of Questions, 1860, PP 1861 [2794] V: 19.

14 Archibald Alison, in First Report from the Select Committee on Combinations of Workmen; Together with the Minutes of Evidence and Appendix, PP 1838 [488], p. 187.

15 Ibid.

16 Ibid.

17 Joseph Kay, *The Social Condition and Education of the People in England and Europe; Shewing the Results of the Primary Schools, and of the Division of Landed Property, in Foreign Countries*, vol. 2 (London: Longman, Brown, Green, and Longmans, 1850), pp. 506–7.

18 Edward William Grinfield, *A Reply to Mr. Brougham's 'Practical Observations Upon the Education of the People; Addressed to the Working Classes and Their Employers'* (London: C. & J. Rivington, 1825), p. iv.

19 Ibid., pp. 10–11.

20 Ibid., p. 17.

21 Ibid., p. 25.

22 Ibid., p. 30.

23 Ibid., p. 30.

24 Evidence from John Dunlop in the Select Committee on Drunkenness, p. 412.

25 Archibald Alison, in Select Committee on Combinations, p. 187.

26 Evidence from Moses Angel of the Jews' Free School, in Popular Education Commission, 2794–V: 48.

27 Evidence by Revd Dr Nathan Marcus Adler, in ibid., p. 19.

28 'The consequences of a scientific education', *Edinburgh Review*, 45:89 (1 December 1826), 194.

29 Revd George Wright, *Mischiefs Exposed. A Letter Addressed to Henry Brougham, Esq., M.P., Showing the Inutility, Absurdity and Impolicy of the Scheme Developed in his 'Practical Observations'* (York: A. Barclay, 1826), p. 16.

30 Elizabeth Neswald, 'Science, sociability, and the improvement of Ireland: the Galway Mechanics' Institute, 1826–51', *British Journal for the History of Science*, 39:4 (2006), 507.

31 Ibid., 531.

32 Martyn Walker, '"Encouragement of sound education amongst the industrial classes": Mechanics' Institutes and Working-Class Membership 1838–1881', *Educational Studies*, 39:2 (2013), 145.

33 Ian Inkster, 'The social context of an educational movement: a revisionist approach to the English mechanics' institutes, 1820–1850', *Oxford Review of Education*, 2:3 (1976), 281.

34 *Annual Report of the Yorkshire Union of Mechanics' Institutes* (Leeds: Yorkshire Union of Mechanics' Institutes, 1859), p. 2.

35 Gerry Wright, 'Discussions of the characteristics of mechanics' institutes in the second half of the nineteenth century: the Bradford example', *Journal of Educational Administration and History*, 33:1 (2001), 14.

36 Walker, '"Encouragement of sound education"', 142–55.

37 Friedrich Engels, *Condition of the Working Class in England* (1844) (London: Panther, 1969), https://www.marxists.org/archive/marx/works/1845/condition-working-class/ch10.htm (accessed 1 August 2021).

38 Steven Shapin, 'A course in the social history of science', *Social Studies of Science*, 10:2 (1980), 246. Also see Steven Shapin and Barry Barnes, 'Science, nature, and control: interpreting mechanics' institutes', *Social Studies of Science*, 7:1 (1977), 31–74.

39 Neswald, 'Science, sociability, and the improvement of Ireland', 503–34.

40 'The Members of the London Mechanics' Institution', *Bell's Life in London and Sporting Chronicle*, 4 December 1825, p. 389.

41 See Joanna Bourke, *Birkbeck: 200 Years of Radical Learning for Working People* (Oxford: Oxford University Press, 2023).

42 Edward Royle, 'Mechanics' institutes and the working classes, 1840–1860', *Historical Journal*, 14:2 (1971), 317.

43 Walker, '"Encouragement of sound education"', 142–55.

44 John Laurent, 'Science, society and politics in late nineteenth-century England: a further look at mechanics' institutes', *Social Studies of Science*, 14:4 (1984), 586.

The founders of the Manchester Mechanics' Institution

Natalie Zacek

As the city of Manchester became an industrial powerhouse in the early decades of the nineteenth century, some of its most eminent inhabitants pondered the question of technical education. In their opinion, the rapid pace of technological change called for a steady supply of scientifically knowledgeable working men. At a meeting on 7 April 1824 at the Bridgewater Arms public house, a small group that included the cotton mill owner Robert Hyde Greg, the scientist John Dalton, and the engineer William Fairbairn decided to establish a training venue modelled on Glasgow's recently established mechanics' institute, which provided evening classes in 'the scientific principles of arts and manufacture'.[1] The Glasgow institute was based on the ideas of Dr George Birkbeck, a lecturer in natural philosophy who had been impressed by the eagerness of the artisans from whom he commissioned scientific instruments to deepen their understanding of the principles of science, and it seemed to these Mancunian scientists and industrialists an ideal model for their project. The Manchester Mechanics' Institution opened on 30 March 1825 in a custom-designed building on Cooper Street, and at its inauguration ceremony Benjamin Heywood, the chairman of its board of directors, invoked the names of renowned pioneers of technology, including James Watt, the creator of the steam engine, and Richard Arkwright, the inventor of the water frame, who had 'raised themselves from humble stations, and who

VI.I The Manchester Mechanics' Institution, Cooper Street, opened in 1825. The institution moved to David Street (now Princess Street) in 1856. Courtesy of Manchester Libraries, Information and Archives, M64350.

have left behind them imperishable monuments of their fame'. Heywood informed his audience that they had the advantage over these heroes because 'every facility of improvement' was now placed in their hands.[2]

No individual did more to establish and maintain the Mechanics' Institution than Heywood, who chaired its board of directors from 1825 to 1840. He and his brother Thomas each donated £500, the equivalent of almost £50,000 in modern money, to its founding, which represented a quarter of its seed capital, and he convinced other members of their extended family to make handsome contributions. Heywood became a 'life member' of the Institution with an initial donation of the current equivalent of a thousand pounds, and on subsequent occasions made similarly large one-off financial gifts to the Institution as well as donating books to its library. He tirelessly sang the Institution's praises, encouraging Mancunians to take pride in this centre of learning and, if possible, become subscribers to it. His support of the Institution was among the accomplishments that gained him both the Fellowship of the Royal Society and the title of baronet, and it is not surprising that it was his portrait that hung in the Institution's reading room.

But it is also Heywood and his family who represent the most obvious connection between the Mechanics' Institution and trans-atlantic slavery. Many of its founders and supporters had prospered from involvement in Manchester's cotton manufacturing industry, which sourced its raw material from slave plantations in the American South, but the Heywood fortune was founded upon participation in the slave trade itself. In 1731 the 14-year-old Arthur Heywood moved from Drogheda in Ireland to Liverpool to begin an apprenticeship with the merchant John Hardman, and in 1739 he married Sarah, the daughter of Hardman's associate Samuel Ogden. In 1741 Arthur's younger brother Benjamin joined him in Liverpool as an apprentice to James Crosby, and would later marry Phoebe, Sarah's sister. These apprenticeships and marriages placed the Heywood brothers at the centre of Liverpool's nascent involvement in the transatlantic slave trade. Their wives' grandfather, John Pemberton, was a key early backer of the trade, investing in his first voyage to Africa as early as 1714. Around the same time that Pemberton was moving out of slave trading, the Heywoods' father-in-law, Samuel Ogden, was entering the business, inheriting his father-in-law's wealth and connections. Arthur's employer Hardman was already involved in slave trading when Arthur began his apprenticeship, and invested alongside Pemberton and/or Ogden in at least 14 slaving voyages, while Benjamin's boss Crosby had by 1745 invested in the first of 23 known voyages.

The Heywood brothers ingratiated themselves with Liverpool's slave trading community at just the right moment. In 1745 they participated as investors in the first of 132 recorded slaving voyages; their network of co-investors in these voyages reads as a Who's Who of Liverpool investors and merchants in the slave trade. For 96 of these 132 voyages information has survived regarding the number of African captives on board, which totals 27,107 individuals. If we apply the average number of enslaved people across those 96 voyages to the whole 132, a rough estimate is approximately 37,000 people. To put this figure into some kind of historical perspective, in the 1790s Liverpool and Manchester were home respectively to 70,000 and 80,000 inhabitants, so the Heywoods transported the equivalent of half of the population of either city from capture in western Africa to enslavement in the Americas.

By the 1770s the Heywoods had amassed such large reserves of capital that they chose to move into banking; in 1773 they set up the Heywood Brothers Bank in Liverpool, which grew so quickly over the next decade that Manchester customers asked for a branch to be opened in their city. The Heywoods opened a Manchester branch, but after a year they found operating two branches inefficient, so in 1788 they dissolved their partnership and Benjamin Heywood took £10,000 of capital and set up his own separate bank, Benjamin Heywood, Sons & Co., bringing his sons Benjamin Arthur and Nathaniel to Manchester as partners. Nathaniel's son was the Benjamin Heywood who became the Institution's great patron. Although neither he nor his father were ever personally involved in the transatlantic slave trade, this commerce was the foundation of the family's wealth, and hence underwrote their support of the Manchester Mechanics' Institution.

* * *

My thanks to Matthew Stallard for his assistance with this research.

Notes

1 W. Johnson, 'The Manchester Mechanics' Institute and the Free Library of Books', *International Journal of Mechanical Engineering and Education*, 34:1 (2006), 4.
2 Benjamin Heywood, *An Address to the Mechanics, Artisans, &c. Delivered at the Opening of the Manchester Mechanics' Institution* (Manchester: Robinson and Bent, 1825), p. 10.

2

Heidelberg in Cottonopolis: how Roscoe brought German ideas to Manchester

Peter J. T. Morris and Peter Reed

Henry Enfield Roscoe (plate 4) was Professor of Chemistry at Owens College between 1857 and 1886, a crucial period in the annals of the college.[1] He not only built up the almost moribund chemistry department into one of the most important in the country, but he also played a crucial role in the college's move to Oxford Road and subsequently in the creation of Victoria University. He was also instrumental in bringing medical education into Owens College. Above all, Roscoe was responsible for the college's assimilation of German ideas regarding the role of science in universities. In this chapter we will examine Roscoe's involvement with Germany, the connections between his chemistry department and Germany, his links with German chemists working in the chemical industry in the Manchester area, and his efforts to introduce German methods into British technical education.

Roscoe and Germany

Roscoe was born in London on 7 January 1833, the son of Henry Roscoe, a lawyer and author, and Maria (née Fletcher). His grandfather was the author, banker and politician William Roscoe of Liverpool.[2] After his father died when he was three, Roscoe was educated at the Liverpool Institute for Boys. He then studied chemistry at University College, London, under two leading chemists, Thomas

Graham (later Master of the Mint) and Alexander Williamson.[3] His cousin Stanley Jevons was a fellow chemistry student.

In the early 1850s it was practically impossible to take a postgraduate degree in England. If a chemist wished to further his education, he had to go to France or Germany. Roscoe decided to go to Heidelberg to study under Robert Bunsen, who had just been appointed to the chair of chemistry in succession to Leopold Gmelin. Roscoe became one of his closest collaborators. We do not know if Roscoe had any affinity with Germany before he went to Heidelberg, and he did not have any family connections with the country. It is also not clear why Roscoe chose Heidelberg, except that Bunsen had a similar approach to chemistry – lying between what we would now call inorganic and physical chemistry – to Roscoe's former teacher Graham. Furthermore, the leading Heidelberg law professor Robert von Mohl was a friend of Roscoe's family.

Whatever his reasons for going to Heidelberg, Roscoe certainly thrived there and quickly gained a PhD in chemistry, which at that time did not require any extended original research; within a decade or so research had become a fundamental part of most German PhDs.[4] He then worked with Bunsen on the photochemical reaction between chlorine and hydrogen, with the aim of creating an actinometer – an instrument that could measure the chemical power of light, specifically sunlight. As this reaction has several complications of which they were only dimly aware, their attempts to develop an actinometer were doomed to failure, but they formulated the Bunsen–Roscoe law of photochemical action. Roscoe also assisted Bunsen with the development of the famous Bunsen burner. Even after he left Heidelberg for good in the autumn of 1856, Roscoe continued to work with Bunsen in the summer vacations until his marriage in 1863. Not only did they continue to work on photochemistry, but Roscoe was also introduced to the atomic spectroscopy developed by Bunsen and the physics professor Gustav Kirchhoff in the summer vacation of 1860. Although he was never directly involved with their research on spectroscopy, Roscoe became a tireless advocate for this powerful new technique, giving a course of lectures on it at the Royal Institution in 1861 and at the Society of Apothecaries in 1868, which was then published as *Spectrum Analysis*.[5] This remained the standard work in English on spectroscopy for many years.

While he was in Heidelberg, Roscoe became friends with several German students and academics, notably Lothar Meyer (now best known for his work on the periodic table), the physicists Hermann Helmholtz and Georg Quincke, the chemists Hermann Kopp and Adolf Baeyer, the Jewish mathematician Leo Königsberger, and the law professor Adolf von Vangerow. Roscoe knew some of them through Bunsen and others through Mohl – Helmholtz was married to Mohl's daughter Anna. Most of these friends were staunch liberals. Mohl had been a member of the revolutionary German parliament of 1848, and even briefly served as minister of justice. Heidelberg was a liberal oasis in the Germany of the 1850s.[6] Thereafter, Roscoe had a rose-tinted view of Germany, until he was disabused of this illusion in the final years of his life, although he was aware of negative aspects of German student life, notably the duelling clubs.[7] In the early twentieth century he became increasingly aware of the risk of war between Germany and Britain. He warned of this risk in a German magazine in 1909 and was greatly saddened by the outbreak of war in August 1914.[8] An even greater blow was the publication of the 'Manifesto of the Ninety-Three' in October of that year, which defended the German invasion of Belgium. To Roscoe's dismay, it was signed by several leading chemists, most notably his friend the physical chemist Wilhelm Ostwald. Thus, in the last year of his life, Roscoe's belief in Germany as a bulwark of civilisation was shattered.

In addition to being an ardent friend of Germany, Roscoe believed strongly in the superiority of the German system of higher education.[9] For the more able students this meant placing research at the heart of the university-level teaching of chemistry, in contrast to the largely examination-based approach used at Oxford and Cambridge. However, research degrees were slow to emerge in Britain, and while Roscoe encouraged his students to carry out research after completing their basic degree, they had to go to Germany if they wanted to take a PhD. For young people working in industry, adopting the German approach implied that the British system of apprentice-ships and night schools would be replaced by more formal teaching in technical colleges. Eventually, after serving on the Samuelson Commission for technical education between 1881 and 1884, Roscoe saw the passage of the Technical Instruction Act by the Conservative government in 1889.[10] This allowed local authorities to support technical colleges, and this was then financed by the so-called whisky money raised by a tax on the sale of alcoholic beverages in public houses. Nevertheless, the British preference for training on the job continued in one form or another until the 1970s. At both educational levels Roscoe insisted on the importance of learning the principles of chemistry before embarking on an industrial career. At the same time he was a pragmatist, and he eventually introduced the teaching of technological chemistry at Owens College to meet the demand from both parents and industrialists (who were indeed often the same people). Hence Roscoe's effort to persuade the British to adopt the German educational system was only partly successful.

The German influence on chemistry at Owens College

Thanks to Roscoe's own education in Germany and his commitment to the German model of scientific education, there were strong links between the Chemistry Department of Owens College and Germany during his tenure as head of department. Roscoe brought his German private assistant William (Wilhelm) Dittmar with him when he took up his chair at Owens College in 1857.

Dittmar left Manchester in 1861, but recommended another German chemist, Carl Schorlemmer (1834–92), as his successor. Schorlemmer was born in Darmstadt, and after overcoming family resistance he set out to become an apothecary, but after attending Bunsen's lectures while in Heidelberg he decided to become a chemist.[11] In May 1859 he began his studies at the University of Giessen under Heinrich Will and Hermann Kopp. In the early 1860s Schorlemmer undertook research on aniline dyes for the Manchester firm of Roberts, Dale & Co. He was then promoted to assistant in the college laboratory (1861), lecturer (1873) and the first professor of organic chemistry in England (1879).

Roscoe and Schorlemmer formed a remarkable duo in the School of Chemistry at Owens College, with Roscoe taking the lead in general and inorganic chemistry, and Schorlemmer in organic chemistry, while they shared the teaching of a course in technological chemistry. According to Roscoe, 'we were very much attached to each other; indeed I do not think during the whole of that time we ever had a disagreement'.[12] The two men co-authored a monumental *Treatise on Chemistry*, Schorlemmer being responsible for the organic chemical and historical sections, and Schorlemmer also published *The Rise and Development of Organic Chemistry* in 1879. He died of lung disease, aged 57, in 1892.

Schorlemmer's interests were much wider than chemistry, and included politics and political philosophy. A member of the German Social Democratic Party (SPD) and the First International, he was a close friend of Friedrich Engels and Karl Marx. Engels had come to Manchester from Germany in 1843 to work in his father's textile works, and it was from his experiences and observations there that he wrote his famous book, *The Condition of the Working Class in England* (English translation, 1887), which reflected his revolutionary republican outlook. In 1864 Engels was appointed to the executive committee of the Schiller Anstalt, which served as the focal point of the German community in Manchester, and as Schorlemmer was a regular musical performer there, this is one place where they are likely to have met and discussed political philosophy, although they first met at the Thatched House tavern in Market Street. Schorlemmer's influence on communist ideology was acknowledged in the Soviet Union and Eastern Europe until the late 1980s. To give a key example, Schorlemmer encouraged Engels's interest in the natural sciences, and gave him the example of alizarin as a synthetic product for the Marxist dialectic of materialism.[13]

Roscoe's subsequent colleagues were all British, except for one colourful figure, William (Wilhelm) Bott, who joined the department in 1883 shortly after taking his PhD at Heidelberg.[14] He became a

2.1 Carl Schorlemmer was an assistant to Roscoe, and then Professor of Organic Chemistry at Owens College from 1874 to 1892. He was a friend of Marx and (especially) Engels, and an important influence on dialectical materialism.

government analyst in the Straits Settlements (now Singapore) in 1890, but moved into mining, setting up a partnership around 1903 using the name William Norman-Bott. While he was mostly active in West Africa, he also argued for the significance of the new Sakhalin oilfields in Russia. When the war broke out in 1914, he was taken to court for failing to register under the Alien Registration Order. Bizarrely, he claimed that he had been born in Manchester, although he was actually born in Wiesbaden in 1861. It was then uncovered that he had been naturalised in 1885, and the case was dismissed. After the First World War, he established a chemical works at Congleton in Derbyshire, where he died in 1935.

The vast majority of students who took chemistry courses in the 1870s and early 1880s were English. There were very few students from Germany, but several from the German community in Manchester. To give just two examples, Alfred Hermann Rademacher was the son of F. Rademacher, the German teacher at the Manchester Commercial Schools, and Augustus Schloesser (and his brothers, who also took chemistry courses) was the son of Reinhold Schloesser, who was born in Neustadt in Prussia. Augustus became an Anglican priest and changed his surname to his mother's maiden name (Mallinson) during the First World War.[15]

Roscoe strongly believed in original research as a key element in a chemical education.[16] In Germany this was easy to achieve, as the main degree was the research-based doctorate of philosophy. However, the PhD degree was unknown in England until the First World War made the hitherto usual trip to Germany to take a PhD unacceptable. Some of the chemists trained by Roscoe took the DSc of the University of London, a research degree but not a common one. Owens College lacked any kind of postgraduate degree, nor did it have anything like Oxford University's research-based Part II (the fourth year of the undergraduate course) which was introduced in 1916. The most promising (or perhaps the most dedicated) undergraduates could carry out research, once they had completed most of the courses, and some of them stayed for up to five years in the laboratory in order to do so.[17] Because the research students at Owens were not registered for a postgraduate degree, we have no official record of their number, but their research output is measurable and gives an indication of how numerous they were. Roscoe records no fewer than 122 papers published by demonstrators and students in his laboratory in the period 1857–87, and 52 of these were published by students, either on their own or with a demonstrator.[18] Roscoe declared that 'I am not going too far when I say that in no laboratory in the kingdom has anything like this amount of sound original work been turned out.'[19]

2.2 The Chemical Laboratory, Owens College: 'in no laboratory in the kingdom has anything like this amount of sound original work been turned out'. Courtesy of the University of Manchester.

Roscoe sent his best students to Germany for training before they returned to Owens College to complete their formation as academic chemists.[20] Needless to say, Heidelberg was the most common destination for these students (chosen by eight of his demonstrators), but Bonn (4) and increasingly Munich (2) were also popular. One would have expected this training abroad to have extended the influence of German chemical education in British universities, but in actual fact only seven out of eighteen of these lecturer-demonstrators stayed in academia. Of these seven, M. M. Pattison Muir at Cambridge University was more important as a writer and lecturer than as a researcher, and Thomas Carnelley, who taught at several colleges, died relatively young. The two standout educators were Tom (later Sir Edward) Thorpe who was at the Normal School of Science (now Imperial College) before becoming the Government Chemist in 1894, and Arthur Smithells, who became pro-vice-chancellor of Leeds University. Julius Cohen worked with Smithells at Leeds and Peter Phillips Bedson went to Armstrong College (later the University of Newcastle).

Roscoe and the German chemistry community in Manchester

A German diaspora settled in Manchester from the early years of the mass industrialisation that had begun in the 1750s, drawn by the entrepreneurial opportunities available as the city grew as an industrial and commercial centre.[21] As Manchester's industry focused increasingly on processing raw cotton, hence its nickname of 'Cottonopolis', the opportunities for calico printers and dyestuffs manufacturers expanded, again attracting a large number of German manufacturers and chemists who already had the training and experience to exploit these new opportunities. Other Germans came to Manchester following the revolutions across Europe in 1848 and 1849, and then in the late 1850s, after opposition to Bismarck's political changes. Manchester's liberal and Nonconformist ethos provided a welcoming environment where individuals and families could thrive.[22] Many in Manchester's German community made major contributions to chemistry and the chemical industry; some remained in Manchester and the surrounding region, taking up academic appointments or starting new industrial entities, while others did the same elsewhere in Britain.

The appointment of Henry Roscoe as Professor of Chemistry at Owens College in 1857, with his connections to German chemists and chemistry, and the subsequent creation of the outstanding School of Chemistry at Owens brought further links with the German community. With his fluent German and his understanding of German culture and history, Roscoe was always a ready host to members of the German community and German visitors to Manchester.

In Manchester there were several meeting places for the German community and for German visitors to the city. These included the Albert Club (1859), the Schiller Anstalt (1860) (to honour the centenary of Friedrich Schiller's birth) and the Thatched House tavern.

The German chemists in the city would also regularly congregate in the chemistry laboratory at Owens College to discuss the latest research and other topics of chemical interest, and Roscoe seems to have welcomed such meetings, which gave him an opportunity to interact when not undertaking other commitments in his busy schedule.

A chemist from a German émigré family who rose to prominence in Manchester before Roscoe's arrival in 1857, though their paths would cross later in many different circumstances, was Edward Schunck (1820–1903).[23] He was born in Manchester, the son of Martin Schunck, an export shipping merchant and later owner of a calico-printing works in Rochdale. Schunck received his early education in Manchester but then studied chemistry at the University of Berlin before being awarded a PhD at the University of Giessen (under Justus Liebig) in 1841. Returning to Britain, he worked for a short period at his father's Rochdale works, before moving to Manchester to become an independent chemical researcher, aiding local manufacturers in developing new dyestuffs and finding reaction pathways that allowed the circumvention of registered patents. Schunck was a founder member of the Chemical Society of London (1841), was elected a fellow of the Royal Society (1850), served as president of the Manchester Literary and Philosophical Society (1882–84), and was an original member of the Society of Chemical Industry (1881). On his death his private laboratory (which was thought to be the finest in Britain) was bequeathed to Victoria University of Manchester, where it formed the Schunck Research Laboratory.

One of the outstanding German organic chemists to visit Manchester was Heinrich Caro (1834–1910). Caro was born in the town of Posen (Poznán in modern-day Poland), a cultural and commercial centre of the Polish-speaking part of Prussia, some 125 miles from Berlin; when Heinrich was 11 his father moved to Berlin. From an early age Caro developed a keen interest in chemistry, completing his secondary education at the Köllnisches Realgymnasium in the autumn of 1852 and gaining entry to Prussia's leading trade school, the Königliches Gewerbeinstitut, while also taking a chemistry course at the University of Berlin. Caro subsequently began an apprenticeship in calico printing, at that time a highly paid prestigious occupation that brought together the chemistry of colours (dyes) with printing processes. Manchester was becoming a major centre for calico printing, and Caro made his first visit to the city in 1857 with the purpose of seeking out the latest techniques for colour dyeing as well as the state-of-the-art printing machinery.[24] Following this successful visit, Caro returned in 1859, and was to spend seven years working for Roberts, Dale & Co. (founded in 1852), with the objective of investigating dyestuff reactions that might aid the company's response to the patent monopoly of its London competitor, Simpson, Maule & Nicholson. Such visits from German chemists resulted in the transfer of knowledge from Germany to Britain and vice versa.[25]

Ludwig Mond (1839–1909) was a German-born and -educated industrial chemist who, having moved to Manchester in 1862, became a close associate of Henry Roscoe, and together they played a major role in the formation of the professional society for those interested in advancing chemical industry. Mond was born in Cassel, and after his early education there he studied under the outstanding chemist Hermann Kolbe at Marburg (1855) and the following year worked with Robert Bunsen at the University of Heidelberg.[26] His main interest was industrial chemistry, but from the standpoint of research and innovation in manufacturing processes. He had become aware of a major problem with the Leblanc process (for converting common salt into alkali) by which one of the by-products of the process, the evil-smelling sulfur waste (which contained most of the expensive sulfur) was dumped on waste ground surrounding the works, or in rivers, canals or the open sea; no attempt had been made by the early 1860s to regenerate the sulfur and thereby improve the overall economics of the process (and reduce its terrible pollution). Mond came to Manchester in 1862 and, having patented a process for treating the sulfur waste, set out to convince manufacturers of the efficiency of his process. The process proved difficult to operate on a large scale but nevertheless was taken up by many companies. In 1873 he became a founding partner (with Sir John Brunner) in the major chemical company Brunner, Mond & Co., with its head-quarters at Winnington in Cheshire.[27]

When serious attempts were made to form a national professional society for those interested in advancing chemical industry, Mond as one of the leading industrialists was very keen to work with Henry Roscoe, because he admired what Roscoe had achieved at Owens College and knew that Roscoe had the standing to bring together both academic and industrial chemists. As a result of their joint efforts the Society of Chemical Industry was formed in 1881.

Ivan Levinstein (1845–1916) was another German-born and -educated industrial chemist who, having moved to Manchester in 1864, established a very large dyestuffs works at Blackley (Manchester).[28] But competition among both British dyestuff manufacturers and continental companies was extremely intense, and Levinstein sought out every possibility for new dyes that would create new market opportunities, and to discover revised reaction pathways for existing dyes to circumvent registered patents. Moving to Manchester, he had expected to find the same research support that German universities provided for industrial companies, but Owens College did not have the level of research staffing that allowed such close collaboration. Roscoe was of the view that the overarching responsibility of Owens was to give its students the fullest knowledge of chemistry, which would later bring long-term benefits no matter the technical challenge faced by a company. This conflict brought Levinstein and Roscoe into regular confrontation at meetings and in journal articles. Levinstein, however, was more successful in his campaign for important changes to the British patent system to protect British commercial interests, and the new Patent Law of 1907 was largely a result of his efforts.[29]

Influence of Germany and German ideas on Roscoe's views on education

From his studies with Robert Bunsen in Heidelberg, Roscoe came to acknowledge that Germany's leading position in science and technology and as a trading nation were linked to its well-funded and well-organised university education system. When in the early 1860s Owens College was becoming overcrowded at its Quay Street site and lacked the space for adequate laboratories and workshops, and for the creation of new faculties that would be a requisite for university status, the trustees decided to review future accommodation needs.[30] Pressed by Roscoe and the college principal Joseph Greenwood, the trustees and the building sub-committee finally agreed that the two should make a tour of German universities (and Zurich University and Technical Polytechnic) during the summer vacation of 1868, so that their findings could be incorporated (where practicable) into plans for relocating Owens College. Their report was submitted to the Owens College building sub-committee and also to the Royal Commission on Scientific Instruction and Advancement of Science when Roscoe and Greenwood gave oral evidence in March 1871.

The German part of the tour took in visits to Bonn, Göttingen, Hanover, Berlin, Leipzig, Freiburg, Heidelberg, Karlsruhe and Munich, and was structured around several key issues including the funding of universities, the provision of laboratories and staffing, the relationship between universities and schools, and the profile of students. While each university had some unique features, Roscoe and Greenwood were able to make some general conclusions in relation to the extension at Owens College. One of their main conclusions concerned the number of teachers at the first level (professors) and second level (lecturers), whereby 'provision is thus made both for the effective instruction of students, and for zealous prosecution of original research'.[31] It was as a result of these staffing levels that it was possible for German university staff to collaborate with industrial companies on specific research topics, such as the development of new dyestuffs. This staff provision was funded through major state financial support; in contrast, in Britain such funding was speculative and with no commitment to year-on-year funding (Owens College received its first government grant in 1890). The German universities had large and well-equipped laboratories that were again funded through the state, while the fees both for lectures and laboratory work were much lower than in Britain. In Germany the costs of equipment and chemicals were paid by the state, whereas at Owens College the professor was responsible for covering these costs.

Roscoe admired the German secondary school 'Gymnasium' system whereby students who had competently completed their studies were awarded the *Abitur* leaving certificate, which became a prerequisite for entry to universities. In contrast, students applying to study at Owens College were too often found not to have the necessary knowledge and understanding to enter on a degree course, forcing the college to impose entrance examinations. Unlike Germany, Britain did not have universal education until the 1870 Elementary Education Act began provision for children aged 5–12 (inclusive) in England and Wales; the age of universal education was extended (as well as its compulsory mandate) by a series of later Acts. Two further issues handicapped Britain and Owens College: public schools and local secondary schools (where they existed) were fee-paying, and attendance at the latter was not compulsory; the curriculum at the secondary level focused on the Classics (Latin and Greek), but from 1840 (and following a series of parliamentary enquiries) the provision of science courses steadily grew through the remainder of the century. It was against this background that Roscoe in his evidence to parliamentary commissions on education emphasised the need for the curriculum (at elementary and secondary levels) to embrace the sciences and for the provision of well-equipped laboratories and the steady supply of trained science teachers. Roscoe was a member of the 1881 Royal Commission on Technical Instruction and took responsibility for investigating its provision in Germany and a number of other continental European countries; for his work and commitment to the Commission and to technical education generally, Roscoe was given a knighthood by the prime minister, William Gladstone, in June 1884.[32]

Unfortunately, the response of successive governments to the recommendations of these commissions was very limited, even when any action was taken. It was not until after his death and well into the next century that Roscoe's vision, based on his visits to Germany and his ongoing correspondence and meetings with German scientists, would begin to be realised.

Roscoe and Germany: the final outcome

Roscoe's attachment to Germany and his high opinion of German scientific education was a constant motif throughout his career, shaping what he thought and did.[33] What was his legacy in this area? At the time of his death the relationship, political and scientific, between Britain and Germany lay in ruins. However, Germany's success in using technology in the First World War could be seen as confirming Roscoe's main argument that German education had prepared the country for the future, while British education had not, an argument put forward in *Nature* just after the end of the war.[34]

While the First World War showed the importance of having a well-trained cadre of scientists and technologists, it also made anything to do with Germany anathema in Roscoe's home country. Indeed, many scientists on the Allied side wanted to leave Germany outside the international scientific community altogether.[35] The delicate task of reintegrating German chemists into the international scientific fold was achieved for chemistry by William Pope. Pope had no connection with Roscoe and was taught by Henry Edward Armstrong at Finsbury Technical College, London.[36] Recently the relationship between Britain and Germany in the field of chemistry has been celebrated by using the German chemists August Wilhelm Hofmann or Justus Liebig as examples of this connection, rather than Roscoe.

Roscoe promoted academic links between Britain and Germany, not least by sending his best students to German universities for their postgraduate training. The First World War ruptured academic and scientific relations with Germany, and thereafter British chemists received their postgraduate education at home. In due course British chemists looked to the United States for their postgraduate training and even their academic careers – the brain drain to America was under way by the 1960s. Germany's former primacy in chemistry gave way to American dominance, and Roscoe's ideal of education based on a German model faded away.

Notes

1 Most of the material in this chapter is covered in detail with further citations in our biography: Peter J. T. Morris and Peter Reed, *Henry Enfield Roscoe: The Campaigning Chemist* (New York: Oxford University Press, 2024). We also cite herein Henry Enfield Roscoe, *The Life and Experiences of Sir Henry Enfield Roscoe DCL, LLD, FRS* (London: Macmillan, 1906), and Sir Edward Thorpe, *The Right Honourable Sir Henry Enfield Roscoe, PC, DCL, FRS: A Biographical Sketch* (London: Longmans, Green, 1916).

2 For the career of William Roscoe and Roscoe's early life, see Morris and Reed, *Roscoe*, ch. 2.

3 Morris and Reed, *Roscoe*, ch. 3 covers his education at UCL.

4 For Roscoe's training at Heidelberg and his research with Bunsen, see Morris and Reed, *Roscoe*, ch. 4.

5 Henry Enfield Roscoe, *Spectrum Analysis. Six Lectures Delivered in 1868 before the Society of Apothecaries of London* (London: Macmillan, 1869).

6 Arleen M. Tuchman, *Science, Medicine and the State in Germany; The Case of Baden, 1815–1871* (New York: Oxford University Press, 1993).

7 Roscoe, *Life*, pp. 62–3.

8 Thorpe, *Sketch*, pp. 179–89.

9 This view is implicit in his speeches and writings; for example, Henry Enfield Roscoe, 'Original research as a means of education', in *Essays and Addresses by Professors and Lecturers of the Owens College, Manchester* (London: Macmillan, 1874), pp. 21–57.

10 For Roscoe's service on the Samuelson Commission, see Morris and Reed, *Roscoe*, pp. 201–16, and for his promotion of technical education while he was an MP, see Morris and Reed, *Roscoe*, pp. 243–5.

11 Karl Heinig, *Carl Schorlemmer, Chemiker und Kommunist ersten Ranges* (Leipzig: BSB B. G. Teubner, 1974); and P. J. Hartog and Anthony S. Travis, 'Schorlemmer, Carl (1834–1892), chemist', *ODNB*, doi: 10.1093/ref:odnb/24831.

12 Roscoe, *Life*, p. 107.

13 O. Theodor Benfey and Anthony S. Travis, 'Carl Schorlemmer: the Red chemist', *Chemistry and Industry*, 15 June 1994, pp. 441–4.

14 For Bott's career, see Morris and Reed, *Roscoe*, pp. 165–6.

15 Information from the class registers in the University archives and an online family tree of the Mallinson family.

16 Roscoe, 'Original research'.

17 Henry E. Roscoe, *Record of Work Done in the Chemical Department of the Owens College, 1857– 1887* (London: Macmillan, 1887), p. 10.

18 Analysis based on Catalogue III, ibid., pp. 45–52.

19 Ibid., p. 17.

20 For Roscoe's students and colleagues, see Morris and Reed, *Roscoe*, pp. 159–72.

21 Su Coates, 'Manchester's German gentlemen: immigrant institutions in a provincial city 1840–1920', *Manchester Region History Review*, 5:2 (1991), 21–30.

22 Mike Rapport, *1848: Year of Revolution* (London: Abacus, 2009), pp. 1–41.

23 T. E. James and Anthony S. Travis, 'Schunck, (Henry) Edward (1820–1903), chemist', *ODNB*, doi: 10.1093/ref:odnb/35974.

24 For Caro, see Carsten Reinhardt and Anthony S. Travis, *Heinrich Caro and the Creation of the Modern Chemical Industry* (Dordrecht: Kluwer, 2000). Their account of Caro's period in Manchester provides valuable insights into Manchester's industrial and scientific circles and its German community.

25 Ibid., p. 49.

26 For Mond, see J. M. Cohen, *Ludwig Mond* (London: Methuen, 1956).

27 Kenneth Warren, *Chemical Foundations. The Alkali Industry in Britain to 1926* (Oxford: Clarendon Press, 1980), pp. 106–20.

28 Anthony S. Travis, 'Levinstein, Ivan (1845–1916), chemical manufacturer', *ODNB*, doi: 10.1093/ref:odnb/45625.

29 Peter Reed, 'The British chemical industry and the indigo trade', *British Journal for the History of Science*, 25:1 (1992), 118.

30 For the extension of Owens College and the creation of the Victoria University, see Morris and Reed, *Roscoe*, chs 5 and 6, and W. H. Chaloner, *The Movement for the Extension of Owens College Manchester 1863–73* (Manchester: Manchester University Press, 1973).

31 Royal Commission on Scientific Instruction, PP 1872 [C 536], p. 502.

32 Roscoe, *Life*, p. 228.

33 This is based on a longer discussion in Morris and Reed, *Roscoe*, pp. 266–8.

34 'War and peace', *Nature*, 102 (14 November 1918), 201–2.

35 Danielle Fauque, 'Reorganizing chemistry after World War I: the birth of the International Union of Pure and Applied Chemistry (IUPAC)', *Rendiconti dell'Accademia Nazionale delle Scienze detta dei XL* (2020), 75–86.

36 For Pope, see C. S. Gibson and K. D. Watson, 'Pope, Sir William Jackson (1870–1939), chemist', *ODNB*, doi: 10.1093/ref:odnb/35575.

3

James Bryce's Manchester: the politics of the remaking of Owens College, 1865–75

Stuart Jones

In May 1920, on the eve of its own centenary, the *Manchester Guardian* marked a special anniversary for the University of Manchester. It was the fiftieth anniversary of the re-foundation – or 'extension' – of Owens College, a process of reform and expansion that led to the college's relocation to the Oxford Road site in Chorlton-on-Medlock. The newspaper carried a lengthy tribute by Viscount Bryce, who was just approaching his 82nd birthday. In 1920 Bryce was a man of world renown as a jurist-historian and academic polymath, a Liberal cabinet minister, an ambassador, a compulsive traveller, a humanitarian campaigner and an internationalist: Europe's leading authority on America, and the world's most prominent campaigner for the Armenians. Bryce wrote evocatively of the men who remade Owens College and so developed the prototype of the civic university in England.[1]

What Bryce did not do was to discuss his own role, beyond remarking that he was one of the few people still living who had known Owens College in its early years. He might have added that one of the others was the *Guardian*'s editor, C. P. Scott, who had assumed the editorship in January 1872, and whose long friendship with Bryce (and with Owens College) was forged in those early days in Manchester.[2] Bryce's connection with the college, and the University that grew out of it, is all but forgotten, and he might seem an unlikely presence in a book that takes its rationale from the

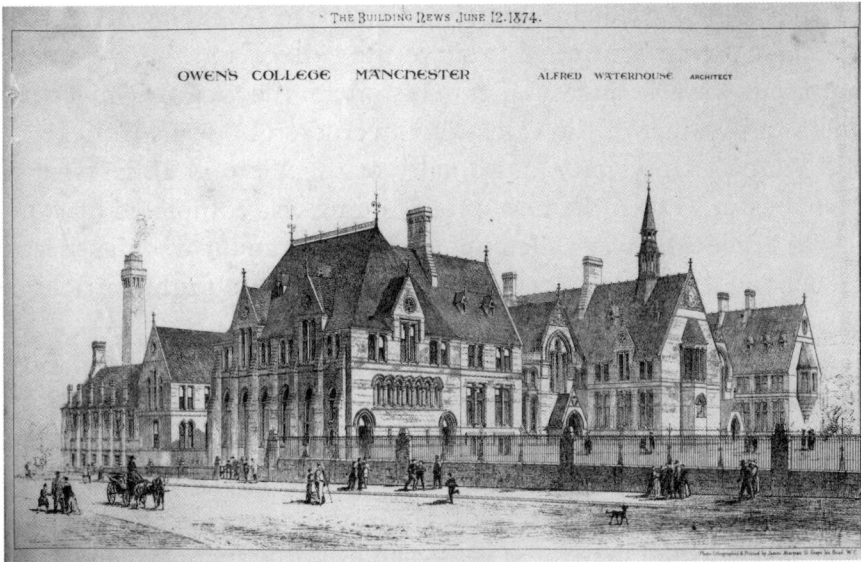

3.1 Alfred Waterhouse's Owens College, opened in October 1873. Courtesy of the University of Manchester.

relationship between ideas and place. A Belfast-born Glaswegian who graduated from Oxford and was based thereafter in London, he never lived in or near Manchester. But the role that he played in the 'extension' movement was an important one, rooted in connections he had already made in the city: Manchester played a key role in his life, and he in its.

Bryce's formal connection is straightforwardly told. He taught law at Owens College from 1867 to 1870, assisting Richard Copley Christie by delivering one course of lectures a year. He then succeeded Christie as Professor of Jurisprudence in 1870, and held the chair until 1875, when pressure of other work forced him to resign. During his brief period in the chair he pioneered a more systematic approach to the teaching of law, one that was designed to meet the needs of practitioners; and in developing a close relationship with the legal profession in Manchester he laid the foundation for a kind of professional education that had hitherto been unknown at Owens. This facilitated the path to the college's assimilation of the Manchester

Royal School of Medicine, which took place in 1872. Having vacated his chair, Bryce continued to serve the college and the university in various ways: as an examiner in history for the Victoria University, and as a member of the Court of Governors of Owens College, of the Victoria University, and finally of the Victoria University of Manchester.[3] When the new Arts Building was completed in 1919, it was Bryce who was called upon to open it; and it was the speech he delivered on that occasion that was the basis for his article in the *Manchester Guardian* a few months later.

This outline, however, misses what might have been his most important contribution. It was he who in 1869 drafted the constitution for Owens College that would serve as the basis for the Owens College Acts of 1870 and 1871 and the reformed college that they established. As this chapter will show, there was a reason why he was asked to do so. The story deserves to be reconstructed, not least because we can trace back to that constitution the structure of university governance that most English universities retained until the end of the twentieth century, and in key respects retain to this day.[4] But the story is not just about constitutions and governance, important as they are: it is also about the relationship between the college and the city. In the extension movement at Owens College we find the origins of the idea of what it means to be a civic university.

'Extension' and its meanings

The significance of the reform of Owens College in 1870–73 is often misunderstood. Even works of some authority attribute changes that occurred then either to an earlier phase (the creation of Owens College in 1851) or to a later one (the formation of the federal Victoria University in 1880).[5] The label 'extension' is confusing, since it seems to imply that what occurred was simply a physical expansion; and it also generates confusion with the later university extension movements, which were really about extra-mural education and other ways of widening access. In 1870–73 there was indeed a physical relocation and expansion, but more importantly 'extension'

entailed a fundamental reconceptualisation of the college's relationship to the public and to the city. The new constitution was important
because it embodied that newly redefined relationship.

John Owens's will had designated his trustees as governors of
the college they were charged with setting up. They would be
responsible for renewing their numbers as vacancies arose. There
was no formal role for the professors in the government of the
college, nor did the trustees have any accountability to the public,
except inasmuch as a general responsibility to the public was implicit
in the nature of a trust: which is to say, not very much at all, since
a trustee's responsibility was normally construed in English law to
be primarily to ensure faithfulness to the founder's intentions. It
is well known that the college struggled in the 1850s, and the
Manchester Guardian famously called it a 'mortifying failure' in
1858.[6] But student numbers picked up in the 1860s, as middle-class
parents started to see the benefits of higher education. It was in
this context that the trustees and the professors began to see the
need for a larger site. A new location would require a much more
substantial endowment, and one that did not impose restrictions
such as those set out by Owens. Re-endowment entailed a closer
relationship with the wider Manchester business and professional
community, rather than just with the friends of John Owens.

The leading figures in the extension movement were the Professor
of Chemistry Henry Roscoe (see Chapter 2), the wealthy cotton
manufacturer Thomas Ashton (1818–98), and the solicitor Robert
Dukinfield Darbishire (1826–1908): all three of them were deeply
embedded in the Unitarian community in Manchester and the region.
In one sense this Unitarian connecdion needs little explanation, for
the Unitarians, though relatively few in number, were influential
among Manchester's mercantile and professional elite, and were
notably committed to educational reform. They were also, as a
denomination, defined by a commitment to freedom of inquiry in
religion, and strongly identified with the cause of non-credal religious
education; and there was a clear affinity between non-sectarianism
and the belief in a properly public form of education open to all.
Something of that attitude can be detected in the response that

Ashton initially gave when Roscoe asked him for help. As Roscoe remembered the exchange, Ashton expressed the view 'that the Governors of the private Trust were strong Churchmen and mainly Tories, with whom he had little sympathy'.[7] He was wrong, in fact, in thinking that the trustees were mostly 'churchmen' (that is, Anglicans), and wrong too in thinking them mostly Tories. He had lumped Owens College together with Manchester Grammar School and Hulme's Charity, both of which were the subject of lengthy contests to break Anglican dominance; and indeed Ashton was closely involved in both reform movements. But the fact that he held this misapprehension is significant: Ashton and many other Dissenting businessmen thought that a private or close trust would tend to be inward-looking and lacking in diversity. It could not be a proper way of governing a genuinely civic or public institution.

So when Ashton came round to Roscoe's plan, it was on the basis that what was entailed was not just a fundraising drive to re-endow the college on a more permanent footing, but also a reconstruction of the form of governance into something appropriate for a college that would *belong* to the city. Only on that basis would it attract the confidence of the business community that would fund the new buildings and the new professorial chairs. The Owens College Extension Act of 1870, which was drafted initially by Darbishire, was unusually explicit on this point.[8] There was a need, it said, to place higher education in Manchester and its neighbourhood 'under the management of a public body rather than of trustees of private nomination'.[9] The instructions issued to Bryce in 1869 when he was asked to draft the constitution made much the same point, drawing attention to 'certain inconveniences inherent in the nature of a private trust' and 'the obvious necessity of affording to the contributing public ... sufficient guarantees for the right use, the due administration and the effective maintenance of the enlarged institution'. These considerations made it necessary to place the new foundation 'on an assured basis as one homogeneous Public Trust'.[10] The constitution as approved in 1870 addressed this issue by establishing a balance in the Court of Governors between a majority of life governors, appointed in the first instance by Act of

3.2 The Owens College Senate in the academic year 1872/73, probably meeting in one of the professors' homes. The principal, Joseph Greenwood, is in the chair on the far right. His successor, Adolphus Ward (then Professor of History and English) is speaking. James Bryce is third from the right. Henry Roscoe is at the front, to Ward's left. Courtesy of Manchester Libraries, Information and Archives, M64044.

Parliament and thereafter self-renewing, and a large minority of governors nominated by or from stakeholder bodies: central and local government, local members of parliament and former students of the college. It also established a balance between the lay governors on the one hand and the professoriat: the former had ultimate responsibility for the good governance of the college, while the latter through the Senate had effective control of academic policy.

The Schools Inquiry and Owens College

If the conclusion were to be drawn that Bryce was simply the instrument of the extension committee, however, that would be a mistake. The committee turned to him for a reason. It was not legal expertise: Bryce was newly called to the bar, and was far less experienced as a lawyer than either Darbishire or Christie, both of them members of the committee. He was not yet Professor of Jurisprudence at the college; neither had he yet been appointed Regius Professor of Civil Law at Oxford, the position he would hold in conjunction with the Owens College chair.[11] Rather, what gave him credibility was the role that had first brought him to Manchester in 1865: he was an assistant commissioner for the Schools Inquiry Commission (the Taunton Commission), and was charged with investigating the secondary schools of Lancashire, including notably its endowed grammar schools. His report had appeared in 1868, so it was still fresh when the Owens College extension committee deliberated: it caused a stir by making a vigorous case for the use of surplus endowments to support the establishment of public day schools for girls, and was also notable for its critique of the inadequate governance of many or most of the schools Bryce inspected.[12]

The problem of endowments was a site of contest in Victorian politics and public policy. Whether founders' intentions should be regarded as sacrosanct was a question of ideological controversy between Liberals and Tories. Liberals, who almost all agreed that founders' intentions could not be regarded as sacrosanct beyond a certain period, were divided over whether endowments were a harmful impediment to the operation of the market or an invaluable provision of property destined to serve public goals, provided that the governance of endowed foundations could be made answerable to public opinion. Bryce took the second position. He conceived of education as a public good, and was concerned not to eradicate educational endowments but instead to subject them to public control so that they served the public interest.[13] His report commended the 'characteristic advantage' possessed by the system of endowed schools, namely its 'publicity': the endowment made the school less dependent on parents, but also more accountable to wider 'public opinion'. Whereas in continental Europe public accountability was believed to require direct state control of schools, the English system of management by local bodies of respectable citizens was 'more conformable to the genius of English institutions'.[14] But Bryce acknowledged that the reality often fell well short of this ideal of accountable local administration. Trustees would sometimes forget the proper conception of trusteeship as 'a delegated authority ... to be exercised for the good and in accordance with the wishes of the people', and instead regard their office as 'a private and personal affair', even 'the hereditary possession of their family or their connexion'. Such 'narrowness and cliqueishness in governing bodies' tended to alienate local citizens from the school. Bryce noted a general wish in Lancashire towns 'to see some measure adopted whch may give a more distinctly public character to the school, and induce the citizens to feel a real practical interest in all that belongs to it'.[15]

Bryce's report found that the narrowness of governing bodies could best be improved by 'the infusion of a representative element'.[16] Later he reiterated the point, recommending 'the reconstitution of governing bodies so as to introduce everywhere a representative element, and to give the townspeople a more direct and lively interest in the welfare of the school'.[17] It was Bryce who formulated the idea of the *representative* governance of a trust, a concept that was introduced into English public discourse by the Schools Inquiry Commission and its aftermath.[18] Representation here did not entail election, however: it was a mechanism for ensuring the responsiveness of governing bodies to a network of stakeholders, typically corporate bodies such as local councils and universities or colleges.

It was, in fact, Bryce's work for the Schools Inquiry Commission that led to his connection with Owens College. Among the first people he met socially when he first visited Manchester were William and Elizabeth Gaskell, to whom he was probably introduced by one of his Oxford mentors, the historian Goldwin Smith. William Gaskell, minister of the famous Unitarian chapel in Cross Street, also taught English literature at Owens College, and knew the college worthies well; his wife, the novelist, died later that year, but the one letter we have from her to Bryce introduces him to Alfred Neild, the calico printer, who had recently succeeded his late father as chairman of the trustees of Owens College.[19]

Moreover, Bryce investigated Owens College in his report, since he was interested in the extent to which the existence of a university college acted as a stimulus to the secondary schools of the city and its region. He thought that Owens should have many more students than it did, and he thought its slow take-off intimately connected with the inadequate state of secondary schooling for boys in Manchester and its hinterland; he also thought that the existence of a thriving university college of the kind Owens had the potential to become would act as a stimulus to the educational ambitions of schoolboys and their parents. He proceeded to set out in his report what was in effect a prospectus for the extension movement. The problem consisted in inadequate buildings located in an obscure part of the city, which meant that the college was largely unknown:

> Lying thus out of sight, the college has in a manner been out of mind also. I found people in Manchester who did not so much as know of its existence, while in some quarters its very newness and unsectarian character seem to have created against it an unworthy and groundless prejudice.

He went on to sketch the solution. Money was needed for lecture rooms and examination rooms, and lodgings were needed to attract students who did not live locally. A change of attitude was needed among employers: if merchants and manufacturers were insistent on getting boys started in warehouses and offices at 15, there was no hope for a university college. But then:

> With these disadvantages removed, with more imposing and commodious buildings, a supply of better prepared candidates, a more assured public position and reputation, it will be able to confer benefits greater than can easily be described upon the education not of Manchester only, but of the whole manufacturing districts in Lancashire, Yorkshire and Cheshire.[20]

Bryce's involvement with Owens College grew directly out of his investigations for the Taunton Commission. Among his enormous collection of papers in the Bodleian Library are some of the notebooks he kept in his work as an assistant commissioner. They include some detailed notes from his interview with Joseph Greenwood, principal of Owens College, probably in March 1866. Then Bryce pivoted straight into an outline of a course of lectures on jurisprudence.[21] He was not lecturing on the subject anywhere at this time;

indeed, he was not called to the bar until June 1867. We cannot be sure whether Greenwood had already broached the subject of Bryce's lecturing for Owens College, but the possibility was certainly in Bryce's mind, and in fact the more elaborate notes he sketched somewhat later in the same notebook correspond closely to the course of seven lectures he gave at Owens in 1871–72.[22]

The alliance of Oxford and Manchester

It is clear that Bryce was known to Owens College through his Schools Inquiry work, and that his report drew him to the attention of the college's extension committee, since he articulated far more explicitly than any of the other assistant commissioners the kind of vision for the proper governance of public educational institutions that they had in mind for Owens College. Just as Bryce thought that Bolton Free Grammar School, to take but one example, needed to be governed in such a way as to give the town as a whole an interest in the school, so the lay leaders of the extension movement envisaged that a remodelled Owens College would be seen as a college belonging to the city: a civic college, though not yet a civic university.

Bryce came to Manchester as a prominent Oxford academic liberal who was committed to political and educational reform. He may well have *chosen* Lancashire as the county he would investigate for the Taunton Commission: certainly he was given the opportunity to express a preference. Goldwin Smith, the *de facto* leader of Oxford liberalism, had called for an alliance of Oxford and Manchester to achieve reform, and the context in which he conceived this alliance was significant.[23] The central aim of the university liberals in the 1860s was to make Oxford and Cambridge genuinely national institutions by abolishing the remaining religious tests that preserved the Anglican monopoly of positions of authority within the universities. Since academic conservatives were increasingly well organised, Smith was pessimistic about the prospect of reform from within, and in any case parliamentary action was needed to repeal the remaining provisions of the Act of Uniformity. So he looked to

Manchester, as a bastion of political liberalism and religious Dissent, to collaborate in an organised campaign for the secularisation of the universities.

Bryce was a key figure in the campaign for the abolition of the tests. Indeed, he personally embodied the cause: coming from a very nonconforming Scots-Ulster Presbyterian background rooted in the Voluntaryist tradition (resolutely opposed to any form of Church establishment), he had contrived to be elected a college fellow at Oxford in defiance of the religious tests, and without conforming. His case was invoked anonymously in the House of Commons: was it not unfair, it was asked, that this talented and multiple prize-winning Presbyterian should have to lose his fellowship when the time came to take his MA degree?[24]

So at the very time that Goldwin Smith advocated an alliance of Oxford and Manchester to press the case for the abolition of the tests, the man who personified the case for abolition was appointed to go to Manchester to investigate the secondary schools of Lancashire. It is unlikely to have been a coincidence, for the year after he first visited Manchester and the month after his interview with the principal of Owens College we find Bryce acting as one of the two organisers of a high-profile public meeting at the Free Trade Hall to generate support for tests abolition. If Bryce was the Oxford half of the alliance of Oxford and Manchester, the Manchester half was none other than Robert Darbishire, the solicitor with whom he would later work on the Owens College extension. It was the tests campaign that first brought the two men together; subsequently Bryce, instructed by Darbishire, drafted the Owens College constitution, and Darbishire drafted the Owens College Bill.[25] The principal speaker at the Free Trade Hall was Dr Frederick Temple, headmaster of Rugby and future Archbishop of Canterbury. Temple was a leading member, perhaps even *the* leading member, of the Taunton Commission, through which Bryce knew him well; and it was Bryce who took advantage of that connection to invite Temple to speak.[26] Bryce was unsuccessful in attracting the Balliol philosopher T. H. Green, an Oxford friend of his and a fellow assistant commissioner on the Taunton Commission, but in the end another Oxford friend, George

3.3 James Bryce in the 1870s, wearing his Oxford DCL gown and hood, a symbol of the struggle against religious tests in universities. Courtesy of the Bodleian Libraries, University of Oxford, MS Bryce 516 fol 9r.

Brodrick of Merton, agreed to come and speak.[27] Brodrick told the meeting that he thought it 'a happy and auspicious circumstance' that the movement that had begun in Oxford should be taken up in Manchester, for it 'above all other places was capable of supplying that motive power without which no movement in this country could be successful', and it was in addition 'the centre of those very classes of England the promoters of this movement were anxious to attract to Oxford'.[28] Temple in particular evidently made an impression on Darbishire, who sent his son Godfrey to Rugby the following year. This marked a significant softening in the attitude of Manchester Unitarianism towards the Anglican public schools: Thomas Ashton also sent his eldest son to Rugby in 1868.[29]

Conclusion

This chapter has used the figure of James Bryce to draw two particular kinds of connection that help us to understand what happened at Owens College in 1870–73. Bryce's report for the Schools Inquiry Commission provided an agenda for the extension movement at Owens, and this insight enables us to see the Owens College extension, and the new constitution that it entailed, as responses to mid-Victorian contests about the endowed institutions, how they should be run and in whose interest. The fact that Bryce studied Lancashire, and that he based himself mostly in Manchester, was important, for the city was the site of highly politicised contests over the control of two notably educational charities, the Manchester Grammar School and Hulme's Charity. Several of the protagonists of the extension campaign were also active in these struggles, Thomas Ashton (later chair of the Hulme trustees) more than anyone; and even before he was personally involved in Owens College he instinctively saw it as raising comparable issues to the struggles over the Grammar School and the Hulme funds.

Manchester had great symbolic importance in Victorian Britain: in 1870 Gladstone called it 'the centre of the modern life of the country',[30] and Bryce and his fellow Oxford Liberals invested it with special significance. It is likely that Bryce chose to investigate Lancashire because it provided him with an opportunity to make allies in the struggle to abolish the remaining religious tests at Oxford and Cambridge, and whether or not it was his intention, it was certainly the outcome. It is indeed a curiosity that the Universities Tests Act of 1871 was passed at the same time as the extension of Owens College was being implemented. The Tests Act had no implications for Owens: its scope was limited to Oxford, Cambridge and Durham, and under the terms of John Owens's will Owens College had been rigorously non-denominational from the outset. But a connection there certainly was. The tests campaign had one big theme: the endowments of the universities could be justified only on the basis that they were of public benefit, and that meant that they should be open to 'the nation', not just to members of one denomination, even if that were the established Church. The abolition of the tests would make Oxford and Cambridge authentically 'national' universities. The extension campaign in Manchester, meanwhile, aimed to give Owens College an institutional structure and a physical presence that would make the Manchester public feel that it was their college. It was to become an authentic civic college, and indeed would soon bid (not immediately successfully) to become a fully fledged civic university.

As for Bryce, after he resigned his chair in 1875 he continued to visit Manchester for a few years as a barrister on the Northern Circuit, but his political career took over, and he sat as Liberal MP for Tower Hamlets (1880–85) and then for Aberdeen South (1885–1907). The connections he made in Manchester were of lasting importance, however, personally as well as politically. In the 1870s he was on intimate terms – indeed, had 'an understanding' with – Meta Gaskell, the daughter of William and Elizabeth, whose house he had visited on his first trip to Manchester in May 1865. This intriguing relationship failed. But when Oxford's Regius Professor of Civil Law eventually got married in July 1889, it was to another Manchester Unitarian, Marion Ashton, the daughter of Thomas Ashton, whose role in the extension movement led him to be hailed as the second founder of the college. The alliance of Oxford and Manchester was complete.

* * *

Thanks are due to Dr Emily Jones for commenting on a draft of this essay. It draws on research undertaken during the tenure of a

Major Research Fellowship MRF-2019-062 awarded by the Leverhulme Trust for the period 2020–23 for a project entitled 'Liberal Worlds: An Intellectual Biography of James Bryce'. I thank the Trust for its support.

Notes

1 Viscount Bryce, O.M., 'The function of the modern university. Some early memories of Owens College and its makers', *Manchester Guardian*, 7 May 1920, pp. 18–19.
2 Bodleian Library MS Bryce 339 (engagement diary for 1873) and 340 (engagement diary for 1875), f. 68, entries for 20 October 1873 and 27 November 1875; MS Bryce 55 f. 57, Harriett Darbishire to Bryce, n.d. [1874]. Henceforth MS Bryce.
3 In fact he hardly attended any meetings of the Court, since he lived in London and latterly in Sussex; but he was valued as a friend of the college/university, and his advice was sought in that capacity.
4 For the enduring legacy of the principle of a lay majority on governing bodies, see 'Hefce criticises self-governance', *Times Higher Education*, 26 March 2009, p. 13.
5 W. R. Ward, *Victorian Oxford* (London: Frank Cass, 1965), p. 159; Graeme C. Moodie and Rowland Eustace, *Power and Authority in British Universities* (Montreal: Queen's University Press, 1974), p. 29.
6 *Manchester Guardian*, 9 July 1858, p. 2; William Whyte, *Redbrick: A Social and Architectural History of Britain's Civic Universities* (Oxford: Oxford University Press, 2015), p. 72.
7 Henry Enfield Roscoe, *The Life and Experiences of Henry Enfield Roscoe, D.C.L., LL.D., F.R.S.* (London: Macmillan, 1906), p. 111. The struggle over the Hulme Trust was very much still in progress at this time.
8 That Darbishire drafted the bill is clear from his correspondence with his son: University of Kentucky Special Collections, Darbishire family papers, Box 1, Folder 14, R. D. Darbishire to Godfrey Darbishire, 5 December 1869.
9 33 & 34 Victoria The Owens College, Manchester, [Ch. 2.] Act, 1870.
10 University of Manchester Archives OCA/7/2/48, 'Manchester and Owens College. Instructions to Mr Bryce to prepare first Sketch of a Constitution'.
11 The one existing study of the extension, W. H. Chaloner, *The Movement for the Extension of Owens College, Manchester, 1863–73* (Manchester: Manchester University Press, 1973), p. 15, is wrong on this point. Chaloner also ignores the constitution.
12 Bryce spoke to the Manchester Ladies' Educational Association in February 1870, on the subject 'On the application of endowments to the education of girls', *Manchester Times*, 5 February 1870, p. 2. In the 1860s and 1870s he was at the forefront of the campaign for girls' and women's education.
13 His fundamental statement on this question was [James Bryce], 'The worth of educational endowments', *Macmillan's Magazine*, 19 (April 1869), 517–24.

This was just before his commission to draft the Owens College constitution. It was a riposte to Robert Lowe, *Middle Class Education: Endowments or Free Trade* (London: Bush, 1868). On this controversy, see H. S. Jones, 'Gladstonian Liberalism, public service and private interests: reforming endowments', in Ian Cawood and Tom Crook (eds), *The Many Lives of Corruption: The Reform of Public Life in Modern Britain* (Manchester: Manchester University Press, 2022), pp. 200–19.

14 Royal Commission to Inquire into Education in Schools in England and Wales (Schools Inquiry Commission), PP 1867–8 [3966] IX: 440 (henceforth cited in the form PP 3966-)

15 Ibid., IX: 442.

16 Ibid., IX: 531.

17 Ibid., IX: 764.

18 The term 'representative governor' (in this sense) does not appear in the Google Books corpus before the 1860s. From the 1870s it becomes common in the context of the constitutions of remodelled governing bodies of schools.

19 MS Bryce 68 ff. 171–2, E. C. Gaskell to James Bryce, 17 May [1865].

20 Schools Inquiry Commission, PP 3966-IX: 721.

21 The interview with Greenwood is noted at MS Bryce 347 ff. 3v–5; the notes headed 'Jurisprudence – Ist Course – Outline' are at ff. 6–7. The contents of the notebook allow us to date the interview to March 1866.

22 The lecture notes are at MS Bryce 347 ff. 44v–87 under the heading 'Jurisprudence – First Course'. The content of the lectures he gave at Owens in 1871–72 can be inferred from the examination paper printed in the Owens College calendar for 1872–73, University of Manchester Archive OCA/4.

23 MS Bryce 16 ff. 12–15, Goldwin Smith to Bryce, 7 July 1869; Christopher Harvie, *The Lights of Liberalism: University Liberals and the Challenge of Democracy 1860–86* (London: Allen Lane, 1976), pp. 84–5.

24 HC Deb 14 June 1865 vol 180 cc200–1 (Göschen) and HC Deb 14 June 1865 vol 180 c245 (Henley, on the other side). Bryce was better informed about the Oriel College statutes than the speaker, George Göschen, and could defer the parting of the ways by taking a BCL rather than an MA as a prelude to an eventual DCL. The DCL would require subscription to the Thirty-Nine Articles (as the MA would have), but by that time Bryce had another ruse.

25 MS Bryce 55, R. D. Darbishire to Bryce, 3 July 1865 and 11 July 1865.

26 MS Bryce 144 f. 9a, Frederick Temple to Bryce, 14 March 1866 [misdated as 1863 by the Library].

27 MS Bryce 73, T. H. Green to Bryce, 23 March 1866. This has been misdated 1868, but the references to the Reform Bill and to a Commons debate on the University Tests Abolition Bill (21 March 1866) make it certain that the year was 1866. For Brodrick, see MS Bryce 42 ff. 44–5, G. C. Brodrick to Bryce, 5 April 1866.

28 'University tests abolition: public meeting in Manchester', *Manchester Guardian*, 9 April 1866, p. 3.

29 *Rugby School Register Volume 2: From 1850 to 1874 Inclusive* (Rugby: Lawrence, 1886). Cf. also University of Kentucky Special Collections Research

Center, Darbishire family papers, Box 1, folder 14, Harriet Darbishire to Godfrey Darbishire, 10 November 1869.

30 Arthur Burns, 'From "Th'Owd Church" to Manchester Cathedral, 1830–1914', in Jeremy Gregory (ed.), *Manchester Cathedral: A History of the Collegiate Church and Cathedral, 1421 to the Present* (Manchester: Manchester University Press, 2021), p. 218.

4

Enriqueta Rylands, founder of the John Rylands Library

Elizabeth Gow

> At the end of the century research in the humanities was brilliantly
> reinforced in an unexpected quarter and by rather unlikely hands.[1]

The 'unexpected quarter' for this new initiative was Manchester,
the 'unlikely hands' those of Enriqueta Rylands (1843–1908), founder
of the John Rylands Library (JRL). That a portion of Manchester's
wealth was philanthropically reinvested in the city was unsurprising.
However, the ambitious scale and scope of the library were unusual.
In under ten years, Rylands established an internationally significant
collection of rare books and manuscripts in one of the world's most
beautiful libraries.[2] The JRL was also notable for its efforts to make
these special collections accessible and useful to people who were
not specialists or established scholars. Rylands's actions transformed
expectations of library philanthropists and rare book libraries and
contributed to Manchester's development as a city of culture and
scholarship.[3] As a woman excluded from academia, Rylands was
'rather unlikely' as a protagonist. Yet her remarkable achievement is
less surprising when considered as part of Manchester's intellectual
history. The JRL was pivotal in her engagement with the world of
ideas, especially her efforts to embed religion in secular education.

Enriqueta Augustina Rylands, née Tennant Dalcour, was born in
Cuba in 1843 and died in Torquay in 1908.[4] She belonged to a
commercial and artistic family that moved from Cuba to New York,
Paris, Liverpool and London. Her father, Stephen Cattley Tennant,

4.1 Portrait of Enriqueta Rylands, the founder of the John Rylands Library. Photograph by Rose K. Durrant. Courtesy of the University of Manchester.

was a Liverpool merchant.[5] Her mother, Camila Dalcour, came from a French-American family who had bought a Cuban sugar plantation with proceeds from land grabs in the American South.[6] As Cuba was under Spanish colonial rule, the Tennant children were baptised into the Roman Catholic Church. Cuba was still a slave society.[7] As well as the Matanzas plantation, the family evidently owned 'house slaves' in Havana.[8] After Stephen died in 1848 and Camila in 1855, the orphaned children were sent to England to stay with members of the staunchly Anglican Tennant family. Sometime in the 1860s – the circumstances are still unknown – Enriqueta Tennant settled in Manchester as a companion to Martha, the second wife of John Rylands (1801–88).[9] A spectacularly successful businessman, John Rylands expanded the small family firm of Rylands and Sons into a corporate empire that dominated the cotton industry and textile trade. He was well-known as a Dissenter – a Protestant Nonconformist. Following Martha's death, John and Enriqueta married on 6 October 1875. Enriqueta adopted her husband's religious affiliations, identifying herself as 'a Dissenter'.[10] Although they were both Congregationalists, the couple advocated for a non-sectarian 'Church Universal'.[11] However, it was Enriqueta who made concrete a comprehensive vision of cultured Nonconformity through the foundation of the JRL.

In the context of a broader neglect of the intellectual contributions of Victorian women, it is unsurprising that Rylands has been marginalised in academic and institutional histories. Rylands herself named the institution in memory of her husband. But this 'act of love' did not make concrete the will and purposes of her husband, as has been assumed.[12] Equally, the significance of Rylands's contribution should not be reduced to her monetary expenditure on books, extraordinary though this was. Rather than remembering Rylands as a 'benefactress' – for her financial gifts – we need to consider the thinking behind them.[13] Recent research on Rylands's library philanthropy has expanded our understanding of her activities as a Nonconformist bibliophile.[14] However, a difficulty persists in perceiving her as a Manchester *mind*. Rylands's approach might be characterised by the motto 'Deeds not Words' – indeed, the motto she chose for the JRL was 'Nihil sine labore' (nothing without labour). As Rylands rarely wrote about or explained her decisions, we must trace her ideas primarily through her actions.

The John Rylands Library

The JRL opened to the public on 1 January 1900. International press coverage immediately recognised the library as a transformational philanthropic act. Rylands's gift was a comprehensive trust encompassing books and building, governance infrastructure, personnel, and investment income. As well as supporting academic research and institutional education, the JRL welcomed the public to exhibitions and lectures. The earliest surviving clue to its origin is an undated letter to her business advisor in which Rylands mentions a 'scheme' to commemorate her husband.[15] She had already commissioned a large, elaborate monument in Manchester's Southern Cemetery.[16] *This* scheme was different. Rylands did not disclose the details, continuing: 'I hope to be able to carry out [the scheme] myself, and if prevented by death from doing so, I will have had opportunities of talking over the matter so that you ... will know in what direction my thoughts run.' Her thoughts became evident in their realisation.

In the autumn of 1889 Rylands purchased land in central Manchester and commissioned plans for a library building from the architect Basil Champneys.[17] Some people thought she was building a theological college. But, before the year was out, she was buying books to fill a library.[18] In 1892 Rylands bought the Althorp Library, an aristocratic collection world-renowned for its early printed books.[19] She integrated this collection with thousands of books and manuscripts acquired separately, including a copy of Shakespeare's first folio and Audubon's massive *Birds of America*. Rylands valued accessibility for a broader audience to the whole collection, including modern books. She said: 'it is my wish that this library shall be of use in the widest sense of the word: for young students as well as for advanced scholars. It is not to be a mere centre for antiquaries and bibliographers, as its rich collection of early printed Books & M.SS. [manuscripts] has led many, I find, to believe.'[20] Rylands's democratising approach – her emphasis on the 'widest' usefulness of the collections – was not without critics.[21] Nevertheless, the new institution was regarded as a model for library philanthropy

4.2 The interior of the John Rylands Library, 1899. Photograph by Bedford, Lemere & Co. Courtesy of the University of Manchester.

in twentieth-century Britain.[22] Rylands's foundation also changed Manchester's reputation on the world stage.

Speculation about Rylands's identity as the anonymous purchaser of Althorp Library coincided with speculation about the place to benefit from the purchase. Would the library go to London, Oxford or Cambridge, or to some 'great centre of population' or 'great modern town'?[23] Modern British cities, like their counterparts on the Continent and across the Atlantic, were characterised by extremes of poverty and wealth and a burgeoning desire (among the wealthy) for cultural prestige.[24] At the inauguration of the JRL, Rylands presented her vision of Manchester: 'And when I use the word city, I use it in the widest possible sense ... as applying to the city in all its manifold activities and life ... literary and educational, mercantile, professional, and industrial, and lastly, what I regard as first in importance, religious.'[25] Again, Rylands took a wide-angle lens, reflecting middle-class pride in commerce and industry allied to education, culture and professionalisation.[26] But she also left her audience in no doubt of the centrality of religion (i.e. Christianity) in her sense of self and city. Rylands was speaking to people who broadly shared her Protestant outlook. Nevertheless, by claiming the pre-eminence of religion, she highlighted a distinction between her project and established civic culture.

The JRL has often been regarded as the result of secular educational philanthropy, but its religious character was broad and pervasive.[27] Giving access to sacred and secular texts in a non-denominational setting mirrored campaigns for non-sectarian Bible education in elementary schools.[28] The library was religious, but like the missions Rylands supported, it was neither narrow, sectarian nor proselytising. The building and collections aligned with the late nineteenth-century historical consciousness that Joshua Bennett has characterised as universalism centred on Christianity.[29] In the JRL, this consciousness was tightly focused on the Bible. Indeed, English Bibles were the prompt for and focus of Rylands's acquisitions of rare books and manuscripts and the library's iconography and exhibitions.[30] The library was a research resource for scholars and a statement about the cultural richness of the communities to whom it was endowed – Manchester and Nonconformity.

Nonconformity and higher education

The explicit yet non-denominational religiosity of the JRL was at odds with existing educational institutions. When Owens College was established, the universities of Oxford, Cambridge and Durham all restricted or excluded Nonconformists. Students and staff faced 'tests' to confirm their Anglican faith. John Owens responded by devoting much of his fortune to higher education 'free from religious tests'.[31] The Nonconformist beliefs and politics of the founders and funders of the college led, counterintuitively, to its secular status. Theology was not taught at Owens but at colleges that trained ministers within different religious denominations.[32] John Rylands supported this distinction by separately funding Nonconformist colleges and scientific or technical education.[33] The 1871 Universities Tests Act abolished religious tests. This enabled the establishment of university colleges for Nonconformists, including Mansfield College at Oxford. Higher education became more accessible. However, because the law permitted religious tests for divinity degrees, academic theology remained divided between the old universities and the disparate denominational colleges. The absence of theology from the curricula of new universities, including at Manchester, undermined claims that they provided a *universal* liberal education. Like her fellow Nonconformists, Rylands wanted a 'pure church' and a secular state.[34] She continued to provide financial support to Owens College as it transformed into the University of Manchester.[35] However, whereas the founders of Owens College excluded theology, Rylands sought ways to bring religion back into education.

Rylands gave her fullest support to initiatives encouraging Nonconformist cooperation, especially in education.[36] Her efforts to embed religion in educational and cultural life reflected her participation in the revivalist Free Church Movement. This interdenominational revival was primarily intended to spread a religious message. However, it also reacted to a perception that Nonconformity was culturally unsophisticated or 'philistine'.[37] The founding of Oxford's Mansfield College was a conscious expression of cultured Nonconformity, dependent on the generosity of donors such as Rylands. Its first principal, A. M. Fairbairn, argued that opening the new buildings with 'no reference to money' would 'crown our churches with dignity, our College with grace, our future with hope'.[38] Significant donations were necessary to avoid the undignified mention of money! In December 1889 Fairbairn suggested that Rylands honour her husband by funding a professorship at his college.[39] Despite sustaining a long-term philanthropic relationship with Mansfield, she preferred to focus on her library project.[40] Indeed, Rylands's largest benefactions and most active involvement were in Manchester, where she could wield personal influence.

In Manchester, the JRL was valued both as a cultural treasury and as a research collection. However, dissatisfaction among academics reflected tensions between these functions. In 1904 Fairbairn noted privately of the JRL that 'the whole department of the history of religion needs carefully to be looked to; modern systematic theology is almost unknown; a great deal of Biblical theology ought also to be added'.[41] This criticism raises doubts about assumptions that Rylands intended to build a theological library.[42] Indeed, she sought advice on acquisitions in various subjects, including foreign history, philology and geography.[43] Although English Bibles prompted Rylands's purchases of rare books, they were part of a collection that tended towards universalism.

In February 1897 Rylands decided the subject of the symbolic statue group in the entrance hall: 'Theology directing the labours of Science and Art'. She had considered the subject carefully, rejecting the architect's proposal for 'Theology, Poetry and Art'.[44] Religion was again central to a broader conversation. The group has been taken as reflecting the contents of the library. However, the figures of art and science do not align easily with the collections as they stood in 1899. The library was humanities-focused, with strengths in history, literature and theology. Neither the physical sciences nor the visual arts were well represented. Instead, the statue group can be understood to stand for Rylands's conception of learning: that the world can be understood (and written about) through scientific methods, artistic interpretation and faith. While scientific methods would usually be associated with the university and faith with the churches, Rylands recognised that these perspectives were relevant to anyone studying any field of concern, from biology to the Bible. In his history of Owens College, Edward Fiddes argued that training for professionals should not be 'merely "practical"' but should 'impart a sound scientific basis of knowledge and a more

4.3 *Theology directing the Labours of Science and Art*, John Rylands Library, 1899. The choice of subject for this statue group in the library's entrance hall was Enriqueta Rylands's own. Photograph by Bedford, Lemere & Co. Courtesy of the University of Manchester.

enlightened outlook'.[45] This 'dual mission' of scientific training and liberal education was characteristic of the civic universities.[46] Rylands believed this to be as relevant to ministers and their congregations as it was to lawyers, engineers or medics. The JRL made knowledge and understanding available to people in Manchester, regardless of institutional belonging.

Nevertheless, the library's relationship with Manchester's educational institutions was crucial to its success and long-term survival. Rylands's religious ideas and personal contacts shaped the foundation of the JRL. The governors she selected also developed the collections. She gave them funds and powers to develop the collections within her scheme. From 1900, Methodist theologian A. S. Peake and Owens historian T. F. Tout led the book committee.[47] Peake annotated his agenda of the first meeting of governors held in December 1899:

> avoid overlapping no law, medicine
> history, theology to be the strong points of the library,
> travels, geographical works.[48]

Peake's sketchy notes outline core collecting areas – history and theology. But they also highlight the importance of the relationship with Owens College. Rylands and the governors intended the JRL to complement rather than compete with the college's library. This complementary collecting strategy would become instrumental in the merger with the University of Manchester Library in 1972.[49] More immediately, it addressed a pressing need at the newly incorporated University.

The Faculty of Theology

On the afternoon of 27 October 1903, Alfred Hopkinson, vice-chancellor of the new University of Manchester, cycled to Rylands's home in Stretford.[50] He had just left a tense meeting. There was a real possibility that the long-cherished idea of setting up a Faculty of Theology might finally come to fruition. The chance was fragile. Some remained vehemently opposed and imposed stringent conditions to stymie the proposal. Hopkinson was a man of action, but why did he rush to Rylands? What assured him of a positive reception from somebody pressed on all sides by demands for charity?[51] Hopkinson later recalled how he made the case to Rylands. He began by recognising her earlier support for Owens College, noting that he had 'not yet asked ... directly for any money'. He then presented the University and the JRL as 'mutually helpful'. Finally, he argued that the faculty would be impossible without funds to 'bring some scholar to Manchester'. Rylands 'at once' promised the money for five years of teaching.[52] Other than a 'strong desire' for anonymity, her only stipulation was that the faculty would 'definitely begin next session'.[53] This critical condition forced the Council to move quickly, countering the opposition's delaying tactics.[54] Historians have argued about whether the origins of the faculty lay in Manchester's denominational colleges or in internal developments at Owens College.[55] However, it was Rylands's response that shifted the balance from doubtful to definite. The shift relied on a 'change in intellectual atmosphere'.[56] Rylands was a crucial but previously forgotten element in this transformation. The movement to treat theology as a 'scientific' discipline and a subject that lent dignity to the new universities was realised in Manchester through Rylands's actions.

The money Rylands promised was used to fund the first two non-denominational chairs of theology in a British university.[57] These chairs in Comparative Religion and Biblical Criticism and Exegesis (analysis of meaning and history) aligned with her interests and purposes.[58] Bible criticism was core: students had to devote four hours a week to the subject for at least two years.[59] Peake, already a JRL governor, was appointed to the chair and became dean of the faculty. Biblical criticism was not only crucial for aspiring ministers. From its start, the faculty also provided public lectures, and it soon broadened its impact by instituting certified courses on religious and biblical knowledge.[60] Intended primarily for Sunday School teachers, these courses also helped to professionalise religious education in schools.[61] Through independence from denominational colleges, the faculty reached different audiences.[62] It belonged to an institution that had admitted women for twenty years (even the medical school had belatedly conceded).[63] Yet the decision to open *theological* degrees to women was unprecedented in the UK.[64] This was a pioneering offer, supported by access to the rich holdings of the JRL. However, women could still be excluded from external lectures, including in denominational colleges.[65] Perhaps stymied by the continued link between theological degrees and professional education, it was not until 1931 that a female student graduated.[66]

The chair in Comparative Religion was also pioneering. Peake argued that 'hesitation' to teach theology at Owens College had been due to a fear of 'sectarian disputes' and a scholarly desire to avoid religious bias.[67] The need for parity was addressed by including comparative religion – namely, the study of 'religion other than Christianity and Judaism' – as a compulsory subject.[68] The developing collections of the JRL, which represented book cultures of religious traditions around the world, underpinned pioneering research.[69] The first professor was T. W. Rhys Davids, a renowned scholar of Buddhism, who later presented his collection of palm-leaf manuscripts to the library.[70] Two other lectureships were also promoted to professorships: Hellenistic Greek and Semitic Languages and Literature. Hope Waddell Hogg and James Hope Moulton, the holders of these posts, were also connected with the JRL.[71] The faculty founders believed that bringing together these subjects alongside comparative religion would enable a 'more objective and scientific view of the subject'.[72] This fitted Rylands's ideas about 'science and art' supporting theology.

Teaching theology at the University posed a risk to Manchester's Nonconformist colleges. Fairbairn, writing from Oxford to Peake – his former student – worried that the move could 'involve the ruin of all the higher teaching in our colleges'.[73] In Manchester, Peake saw things differently. He argued that the number of theological colleges made Manchester a 'peculiarly favourable field for the experiment'.[74] The difficulty in bringing together these disparate voices was solved, in part, by Rylands. In addition to her financial contribution and the library's holdings, she supplied a model of cross-sectoral governance. The library's governance linked the University to the 'manifold activities of the city', including the theological colleges.[75] From Owens College to the National Council of Free Churches, these organisations reflected Rylands's broader philanthropic priorities: education, Dissent and Manchester.[76] Peake, experiencing Rylands's endeavours first-hand, was surprised by the degree of 'harmonious co-operation'.[77] Indeed, he regarded the success of this 'experiment' as critical to the success of religion in an increasingly secular society. Without mentioning Rylands, John Rogerson tentatively suggested that 'the beginnings of the ecumenical movement' influenced the faculty's founding.[78] By putting Rylands back in the picture, the significance of religious ecumenism in Manchester's intellectual life becomes clear.

Recognition

Fiddes regarded the JRL as Rylands's greatest 'service to the world of scholarship'.[79] During her lifetime, Rylands was celebrated for her philanthropy. Most famously, she was the first woman to be given the Freedom of the City of Manchester. One English schoolchild aspired to 'give gifts to my country, like Mrs. Rylands'.[80] A scroll and silver casket presented along with the Freedom of the City commemorated Rylands's civic honour. However, her contributions to the world of ideas are less visible. When Owens College awarded Rylands an honorary doctorate in 1902, its representative recognised her 'splendid munificence' in founding and 'lavishly' endowing the JRL. Yet he also recognised her 'far-sighted sagacity' in writing the library's rules of government and her 'watchful and discriminating generosity' in developing its collections.[81] Rylands's financial generosity, political acumen and collecting legacy were similarly critical to Manchester's Faculty of Theology. In 1910 the University renamed the chair of Biblical Criticism in her memory.[82] Yet the Rylands Professorship was presumed to refer to John Rylands as much as to his widow.[83] Rylands's treasured doctoral gown was offered to the JRL, but there is no record of the librarian's response, which was presumably negative.[84]

Presenting Rylands's contributions to theological education alongside her library philanthropy brings her religious ideas into view. Her wide perspective on religion and education lies behind the John Rylands Library and its transformative effect on Manchester's cultural identity. Her multi-modal enactment of her religious values also contributed to the democratisation of higher education in the twentieth century. Seeing Rylands as a 'Manchester mind' reveals her importance to a broader transformation of intellectual life in Britain.

Notes

1 David Owen, *English Philanthropy, 1660–1960* (Oxford: Oxford University Press, 1965), p. 370.
2 Jacques Bosser, *The Most Beautiful Libraries of the World* (London: Thames and Hudson, 2003).
3 Elizabeth Gow, 'Enriqueta Rylands: The Public and Private Collecting of a Nonconformist Bibliophile, 1889–1908', PhD thesis, University of Manchester, 2023, https://research.manchester.ac.uk/en/studentTheses/enriqueta-rylands-the-public-and-private-collecting-of-a-nonconfo (accessed 28 March 2024).
4 Douglas A. Farnie, 'Enriqueta Augustina Rylands (1843–1908), founder of the John Rylands Library', *BJRL*, 71:2 (1989), 3–38.
5 Catherine Davies, 'Stephen Cattley Tennant, 1800–48', *BJRL*, 85:2 (2003), 115–20, doi: 10.7227/BJRL.85.1.7.
6 Raul Ruiz, 'Mrs Rylands's Cuban origins', *BJRL*, 85:1 (2003), 121–26, doi: 10.7227/BJRL.85.1.8; Blake A. Watson, 'Buying West Florida from the Indians: the Forbes Purchase and Mitchel v. United States (1835)', *FIU Law Review*, 9.2 (2014), doi: 10.25148/lawrev.9.2.13.
7 Josep Maria Fradera and Christopher Schmidt-Nowara, *Slavery and Anti-slavery in Spain's Atlantic Empire* (New York: Berghahn Books, 2013).
8 For a fuller discussion of the place of slavery in Enriqueta Tennant's family, see my blog post at https://rylandscollections.com/2020/09/14/whiter-than-white-enriqueta-rylands-cuban-roots/ (accessed 28 March 2024).
9 Douglas A. Farnie, *John Rylands of Manchester* (Manchester: John Rylands University Library of Manchester, 1993).
10 JRL Archive, JRL/1/6/1/1, letter from Rylands to her business advisor William Linnell, 13 February 1893.
11 Elizabeth Gow, '"Not slothful in business": Enriqueta Rylands and the John Rylands Library', in Clyde Binfield, G. M. Ditchfield and David L. Wykes (eds), *Protestant Dissent and Philanthropy in Britain, 1660–1914* (Woodbridge: Boydell and Brewer, 2020), pp. 205–22.
12 Edward Fiddes, *Chapters in the History of Owens College and of Manchester University, 1851–1914* (Manchester: Manchester University Press, 1937), p. 187.
13 Anthony Hobson, *Great Libraries* (London: Weidenfeld and Nicolson, 1970), p. 268.
14 Gow, 'Enriqueta Rylands'.
15 JRL Archive, JRL/6/1/3/6, letter from Rylands to Linnell, dated 'Sunday' [June or July 1889].
16 'Monument to the Late Mr John Rylands', *Manchester Guardian*, 1 August 1890, p. 5.

17 JRL Archive, JRL/5/2/1/1, copy letter book, 1889–1895, p. 3. See John Hodgson, 'Carven stone and blazoned pane: the design and construction of the John Rylands Library', *BJRL*, 89:1 (2012), 19–81, doi: 10.7227/BJRL. 89.1.3.

18 'Manchester City Council', *Manchester Courier and Lancashire General Advertiser*, 5 February 1891, p. 3; JRL Archive, JRL/6/1/1/7/1, Richard D. Dickinson, invoice, 29 November 1889.

19 Peter H. Reid, '"The finest private library in Europe": a brief study of the bibliophile Spencers of Althorp', *Library History*, 14:1 (1998), 65–71, doi: 10.1179/lib.1998.14.1.65.

20 JRL Archive, JRL/6/1/3/6, letter from Rylands to Linnell, 13 April 1896.

21 For example, in 'The jubilee of Owens College', *Athenaeum*, 22 March 1902, pp. 371–2.

22 Christine Alexander, 'Lord Brotherton', in William Baker and Kenneth Womack (eds), *Nineteenth-Century British Book-Collectors and Bibliographers* (London: Gale, 1997), pp. 37–45.

23 'Sale of the Althorp Library', *The Times*, 29 July 1892, p. 7.

24 James Moore, *High Culture and Tall Chimneys: Art Institutions and Urban Society in Lancashire, 1780–1914* (Manchester: Manchester University Press, 2018); Paul Dimaggio, 'Cultural entrepreneurship in nineteenth-century Boston: the creation of an organizational base for high culture in America', *Media, Culture & Society*, 4:1 (1982), 33–50, doi: 10.1177/016344378200400104.

25 'The John Rylands Library', *Manchester Guardian*, 7 October 1899, pp. 6–7.

26 Simon Gunn, *The Public Culture of the Victorian Middle Class: Ritual and Authority and the English Industrial City, 1840–1914* (Manchester: Manchester University Press, 2000).

27 Gow, '"Not slothful in business"', pp. 218–21.

28 Ibid., pp. 216–17.

29 Joshua Bennett, *God and Progress: Religion and History in British Intellectual Culture, 1845–1914* (Oxford: Oxford University Press, 2019), pp. 2–8.

30 Gow, 'Enriqueta Rylands', pp. 198–230.

31 Fiddes, *Chapters in the History*, p. 14.

32 Ibid., pp. 14–19.

33 Joseph Thompson, *The Owens College: Its Foundation and Growth; and Its Connection with the Victoria University, Manchester* (Manchester: J. E. Cornish, 1886), pp. 460, 641, 646.

34 Timothy Larsen, *Contested Christianity: The Political and Social Context of Victorian Theology* (Waco, TX: Baylor University Press, 2004), p. 156.

35 Fiddes, *Chapters in the History*, pp. 185–7.

36 Gow, '"Not slothful in business"'.

37 Michael R. Watts, *The Dissenters: The Crisis and Conscience of Nonconformity* (Oxford: Clarendon Press, 2015), pp. 181–90.

38 A. M. Fairbairn, 'Our work and its results', undated circular, Mansfield College Archives.

39 W. B. Selbie, *The Life of Andrew Martin Fairbairn* (London: Hodder and Stoughton, 1914), pp. 240–3.
40 Gow, '"Not slothful in business"', pp. 213–14.
41 UML, Peake Papers IV/368, letter from Fairbairn to Peake, 1 April 1904.
42 Gow, 'Enriqueta Rylands', pp. 82–3.
43 JRL Archive, JRL/6/1/2, letters to Rylands from J. Arnold Green, April–October 1894.
44 Hodgson, 'Carven stone and blazoned pane', 52.
45 Fiddes, *Chapters in the History*, p. 112.
46 Sarah V. Barnes, 'England's civic universities and the triumph of the Oxbridge ideal', *History of Education Quarterly*, 36:3 (1996), 271–305, doi: 10.2307/369389.
47 V. H. Galbraith and (revised by) Peter R. H. Slee, 'Tout, Thomas Frederick (1855–1929), Historian', *ODNB*, doi: 10.1093/ref:odnb/36539; Timothy Larsen, 'A.S. Peake, the Free Churches and modern biblical criticism', *BJRL*, 86:3 (2004), 23–54, doi: 10.7227/BJRL.86.3.3.
48 Peake Papers, IV 1580, Peake, notes on the first meeting of JRL Council of Governors, 11 December 1899.
49 'Notes and news', *BJRL*, 55:1 (1972), 1–3.
50 Alfred Hopkinson, 'The Theological Faculty: how it came into existence', *Manchester Guardian*, 9 October 1929, p. 5.
51 Gow, '"Not slothful in business"', p. 211.
52 Hopkinson, 'The Theological Faculty'.
53 UML, Owens College Archive, OCA/9/1/10, Owens College, Minutes of Proceedings of Council, 28 October 1903, pp. 84–5.
54 C. Lees and A. Robertson, 'Community access to Owens College, Manchester: a neglected aspect of university history', *BJRL*, 80:1 (1998), 125–52 (p. 150), citing H. B. Charlton, *Portrait of a University, 1851–1951* (Manchester: Manchester University Press, 1951), pp. 114–17.
55 Ronald H. Preston, 'The Faculty of Theology in the University of Manchester: the first seventy-five years', *BJRL*, 63:2 (1981), 463–84; John William Rogerson, 'The Manchester Faculty of Theology 1904: beginnings and background', *BJRL*, 86:3 (2004), 9.
56 Fiddes, *Chapters in the History*, p. 24.
57 Scott Mandelbrote and Michael Ledger-Lomas (eds), *Dissent and the Bible* (Oxford: Oxford University Press, 2013), p. 32.
58 Larsen, 'Biblical scholarship in the twentieth century: the Rylands Chair of Biblical Criticism and Exegesis at the University of Manchester, 1904–2004', *BJRL*, 86:3 (2004), 6.
59 UML, FTH/1/1, University of Manchester, Faculty of Theology minutes, 25 May 1904.
60 Ibid., 14 and 27 July 1904 (lectures), 28 November 1905 and 8 February 1906 (courses).
61 David William Bebbington, 'Gospel and culture in Victorian Nonconformity', in Jane Shaw and Alan Kreider (eds), *Culture and the Nonconformist Tradition* (Cardiff: University of Wales Press, 1999), p. 52.
62 Lees and Robertson, 'Community access to Owens College', 147–8.

63 Joanne Young, '"Amongst stuffed beasts and fire-buckets": women and university spaces at Owens College, Manchester 1883–1900', *BJRL*, 96:2 (2020), 98–101.

64 Alan P. F. Sell, *The Theological Education of the Ministry: Soundings in the British Reformed and Dissenting Traditions* (Eugene, OR: Pickwick, 2013), p. 268.

65 Faculty of Theology minutes, 27 July 1904.

66 Sell, *The Theological Education of the Ministry*, p. 268.

67 Peake Papers XIX/65, Peake, draft article, undated.

68 Fiddes, *Chapters in the History*, p. 156.

69 John Hodgson (ed.), *Riches of the Rylands: The Special Collections of the University of Manchester Library* (Manchester: Manchester University Press, 2015), pp. 149–67, 169–85.

70 N. A. Jayawickrama, 'Pali manuscripts in the John Rylands University Library of Manchester', *BJRL*, 55:1 (1972–73), 146–76.

71 Preston, 'The Faculty of Theology', 4–5.

72 Fiddes, *Chapters in the History*, p. 157.

73 Peake Papers IV/368, letter from Fairbairn to Peake, 1 April 1904. The MS letter has 'the higher' added in superscript as an afterthought.

74 Peake Papers III/560b, Peake, 'Statement on founding of Faculty of Theology', annotated typescript c. 1920.

75 'The John Rylands Library', *Manchester Guardian*, 7 October 1899, pp. 6–7.

76 Enriqueta Rylands, *The John Rylands Library: Statement of the Constitution* (Manchester, 1900).

77 Peake, 'Statement on founding of Faculty of Theology'.

78 Rogerson, 'The Manchester Faculty of Theology', 19.

79 Fiddes, *Chapters in the History*, p. 187.

80 'School children's ideals', *The Woman's Journal*, 12 May 1900, p. 145.

81 Josephine Laidler and Owens College, *Record of the Jubilee Celebrations at Owens College, Manchester* (Manchester: Sherratt and Hughes, 1902), p. 86, quoting Professor A. S. Wilkins's presentation.

82 Peake Papers VIIIA/48, letter from Fiddes to Peake, 24 March 1910.

83 Larsen, 'Biblical scholarship', p. 6.

84 JRL Archive, JRL/4/1/1928/Beer, letter from Lucy Beer to JRL Librarian Henry Guppy, 13 September 1928.

5

William Boyd Dawkins: race, geology and the deep past in Manchester, 1869–1929

Chris Manias

Modern universities divide ways of knowing the past according to time, sources, scale and the presence of humans. History, archaeology, evolutionary biology, geology and cosmology are located in different departments. Indeed, at the modern University of Manchester these subjects are divided across all three faculties, and at least five departments and subject areas. With current vogues for interdisciplinarity, some are attempting to overcome these barriers, from various perspectives. There are drives for 'big history', tracing long developments from the formation of the universe and the earth, to the development of life, and through human societies of changing levels of complexity. As Ian Hesketh has noted, these themselves are based on much longer traditions of universal history, and are often quite unreflective on this background.[1] There has also been a desire to connect the geological and the human in more critical veins. Debates over the 'Anthropocene' rest on trying to think about humans in geological timeframes and scales. Our current attempts to resolve these differences date back to the formation of these disciplines and ways of knowing. And here, a particularly important – although also complex – role was played by a figure who made a strong institutional home at Manchester.

William Boyd Dawkins was the first Professor of Geology at Owens College and worked across a range of fields, in public, academic and economic debate. He wrote popular and scholarly

5.1 William Boyd Dawkins. Courtesy of the University of Manchester.

works reflecting on the depths of prehistory, surveyed landforms and geological formations for coal and infrastructural possibilities, and created educational and public institutions directed towards knowledge of the past. These were united by a vision graded between different modes of knowing the past – the geological, the archaeological and the historical – all subjects that had themselves transformed across the nineteenth century. He saw these fields as connected, like rock strata in a layered landscape. In his 1880 work, *Early Man in Britain*, he wrote:

> Of the many fields of inquiry opened out by the intense mental activity of this century, there is none which promises to be more fruitful than that which has been won by the joint labours of the geologist, the student of prehistoric archaeology and the historian.[2]

Each added to understandings of time and the past – the geologist 'tells of continents submerged, and of ocean bottoms lifted up to become mountains; and he points out to us that side by side with the ever-changing conditions of life there were corresponding changes in living forms'. Next the archaeologists 'have raised the study of antiquities to the rank of a science by the use of a purely inductive method', to show 'the steps by which man slowly freed himself from the bondage of the natural conditions under which all other creatures live'. And finally, historians showed humans 'in the high state of civilisation marked by the use of letters'.[3] These were three interconnected means of knowing the past, which together transformed understandings of nature and humanity.

Dawkins's career connected the worlds of scholarly associations, localised excavations and economic expansion. After a childhood in Wales, Lancashire and Somerset, he studied natural sciences at Oxford and worked at the British Geological Survey. In 1869 the trustees of Owens College appointed Dawkins the curator of the museum collections they had just purchased from the Manchester Natural History Society. This appointment was apparently recommended by the scientific naturalist Thomas Henry Huxley, who had supervised Dawkins at the Jermyn Street Museum and Royal College of Mines in London. From Huxley's perspective, supporting Manchester collections would help break out of the strictures of the London scholarly establishment, with Manchester as an ideal place to promote new professional and technical institutions. Dawkins was made a lecturer in geology in 1872 and professor in

1874, as part of Owens College's aim to expand its teaching provision in geology. This cemented his role as a key figure in Manchester's scholarly and intellectual community, and enabled him to develop a position as one of the core authorities on the ancient past of Britain. In Manchester, Dawkins was strongly involved in working-class education, and supported the entry of women into university education from as early as the 1870s. Indeed, Marie Stopes, who had a career as a palaeobotanist at Owens College prior to her much better-known later activity as a birth control pioneer, sexologist and eugenicist, recalled how her initial appointment was almost blocked by the University Senate, but was eventually ratified when 'dear Old Boyd-Dawkins [sic] got on the war path and fought valiantly for me'.[4] Dawkins was also a leading figure in the Royal Society, the Geological Society of London, the Society of Antiquaries and the British Association for the Advancement of Science. He mixed close involvement with national scientific institutions with a strong power-base in Manchester, and links across the countryside of northern England and Wales. Beyond this, a strong global and imperial eye gave him a position in relation to the exercise of British power around the world, and all the ideologies that that implied.

This chapter will follow Dawkins's work to think about the significance of different means of knowing the past in the Victorian period, and some of their implications today. Looking at Dawkins should not be a simple story of the triumph of a particular brand of cross-disciplinary work, and of how knowledge of the natural world can connect with humans. In many respects, his career shows some of the difficulties and tensions in these processes, as well as several darker sides. His work on geology depended on the extraction of fossil fuels, and the growth of coal-based and imperial infrastructure. He amassed a huge amount of material in the north-west of England, often through muscling in on sites already being prospected by other scholars. And his connection of humans and nature was based on Victorian preoccupations with race and civilisation. We will move through his role in relation to the geological past, archaeology and his position in history, and think about how these graded into one another, and, moreover, how the deep past became a tool of control and mastery of the world.

Geology

Geology was one of the defining sciences of the nineteenth century. As the use of steam power increased, the search for coal deposits was a key economic and technical driver – and one that gave huge economic importance to coal-producing regions. Iron, copper and tin for manufacturing, and clay and gravel for building materials, were also crucial geological products. For all these sectors, knowledge of the Earth was critical. And geology was not just economically useful, but culturally and intellectually dramatic. As geologists mapped and surveyed the rock strata, they identified that differing layers of rock corresponded to different periods of the Earth's history (and could be read like 'the pages of a book', according to a widespread metaphor – admittedly with numerous gaps and omissions). They also contained fossils of long-dead animals, illustrating a series of lost worlds before human existence. There is a tendency in much writing on geology to separate 'economic' geology, engaged in the search for mineral resources, and the more conceptual sides of the field. And indeed, as noted by Geoffrey Tweedale, much early history of geology focused on its intellectual importance at the expense of examining the crucial economic elements.[5] This was an important feature of geology in terms of its self-presentation, but of course it obscures how deeply the search for minerals and the construction of a history for the Earth and life were connected. Geology transformed the economy, the map of the current environment and understandings of time and nature, linking the products of the Earth with former eras of its history.

Dawkins's career intermixed these two elements of geology. In his early career, he often secured important fossil sites for himself in a manner similar to that used by coal prospectors to secure mineral resources. Initially in the 1860s he worked at the 'Hyena Den' at Wookey Hole in Somerset, near his family home at Westonzoyland, which contained the remains of cave hyenas and some rare human artefacts. After being appointed at Manchester, he spent many summers in the caves of the Severn Valley, Derbyshire and North Wales. And from the 1870s he became a leading figure in the excavation of Creswell Crags in Derbyshire, a site found by lime quarriers and initially excavated by local antiquarians Thomas Heath and John Magens Mello. However, Dawkins used his institutional power and support from luminaries such as John Lubbock to control the excavations. He was a combative figure, and clashed with Heath over the authenticity of key finds from the site: a carved image of a reindeer and a tooth of Machairodus, a sabre-toothed cat.[6]

Dawkins's economic consulting ramped up from the 1880s to the 1920s. He was involved in a range of projects, including advising on water supply across the country, and late in his career he started courses at Manchester for mining engineers. He also provided geological advice for the construction of the Manchester Ship Canal, and was a prime advocate for its construction. Meanwhile, in the south-east of England, he provided geological and engineering advice on plans for a tunnel under the English Channel from the early 1880s. While several trial tunnels were dug, the project was eventually cancelled owing to costs and government and military suspicion of the idea. The surveying also led to examinations of the geology of Kent and the location of a large coalfield.[7] These economic elements of Dawkins's career were a highly complementary strand to his more theoretical and conceptual work on the history of life.

In Dawkins's works and writings, he tended to focus on two major geological periods. First, the Carboniferous, an ancient period with strange fossil plants and an apparently tropical environment. It was doubly important as the era in which most of Britain's coal was formed, and Carboniferous deposits were particularly widespread in the north of England and Wales – and Naomi Yuval-Naeh has written on how coal connected with the fascination for deep time and exotic plants in the Victorian period.[8] The ancient land was reconstructed to show huge shifts in landscape in the depths of prehistory. In one lecture to the British Association, Dawkins conjectured about the existence at this time of a 'great north-western continent' named 'Archaia', 'which occupied the area of the North

Atlantic in the direction of Iceland, Greenland, and a large portion of North America, [and] extended southwards' to reach what was now northern Britain at its fringe coastal regions. This land was 'diversified by chains of mighty lakes, embosomed in luxuriant forests of conifers, and various Lepidondendron and Calamitean trees'. Over the ages, this coast transformed into 'a delta of a mighty river analogous in every particular to the Mississippi', and the islands in this delta formed the British coal measures, which were now being mined to drive the steam-power economy.[9]

This was a colossal history, with huge forces operating on a vast timescale. But the Carboniferous was also a more local affair. In the 'Science Lectures for the People' series, Dawkins spoke 'On Coal' for an audience of working-class Mancunians. This lecture clearly positioned the importance of the commodity on a local and national level. Dawkins began:

> I think that we who live in Manchester have an especial right to know what coal is. The very fact that Manchester stands where it is, and the very fact that we are living in a city of over 500,000 souls, instead of in a little village, is simply owing to the circumstance of coal being found here. Coal is the great centre of our prosperity, and upon it depends nearly all the success of our manufacturing enterprise. The political economist will tell you that upon it the future of England mainly depends.[10]

The rest of the lecture went through the structure of coal, fossils of seeds and club moss often found within it, the location of coal deposits throughout Britain, and the ancient Carboniferous forests where it had been formed. And Dawkins closed his lecture by noting that most of the samples that he had discussed had been collected by 'two working men in Oldham' named as the millworkers Mr Butterworth and Mr Whittaker,

> by using that common sense which all of you must possess, and which all of you can apply, if you only care to look outside yourselves, and to look into nature, and to see what wonders and what mysteries there are lying at your very doors. I am quite sure that none of you would be at all less happy, but far more happy for any study whatever that would take you out of your own ruts of life ... and show you what beautiful things there are in nature.[11]

This patronising statement of course belied the fact that people like Butterworth and Whittaker already knew considerable amounts about coal, and that the cultures of working-class education in which Dawkins was (in his own patrician way) participating were already engaged in these issues.

While coal was significant in the drama of Earth's history and the carbon-based economy, most of Dawkins's attention focused on more recent periods, especially the Pleistocene, the geological era immediately before the present, which in these years was being defined as the 'Ice Age' (or – according to some, such as Dawkins's rival James Geikie – multiple 'Ice Ages'). The fossils showed a fauna of mammoths, woolly rhinoceros, cave bear, hyena, lion, reindeer and horse, existing in a land of harsh cold and large glaciers. Many of the fossils were found in caves (where they had escaped the scouring of the glaciers), and gave a sense of strangeness and mystery.

The world of the Ice Age was visually reconstructed in a sketch produced by the artist M. Rowe in a private commission for Dawkins, and now kept in the archives of the Manchester Museum. It depicted 'The Severn Valley in the Pleistocene Age', this region being one that Dawkins had particularly studied, excavating fossils and taking the remains back to his Manchester collections. It showed the more striking Pleistocene beasts, including deer, bison, two species of fossil elephant and hippos (whose fossils were found as far north as Yorkshire). In the background were the ferocious British lions, hunting wild horses and – even more dramatically – early humans, naked and tending fires in front of caves presumably being used as habitation.

In this image, and in his wider popular works, *Cave Hunting* and *Early Man in Britain*, Dawkins's attention focused on the fauna, in its ferocity and strangeness. However, the most significant element of the landscape, and the greatest mystery to solve, was the presence of humans. The question of 'human antiquity', and when humans had appeared in the chronology of life, was a live one in the mid-Victorian period. While a series of finds in Britain and France were publicised around 1859–60 to assert that humans had existed within geological periods and alongside the cave bears and mammoths

5.2 *The Severn Valley in the Pleistocene Age*, a sketch commissioned by Dawkins in the 1870s. Courtesy of the Manchester Museum.

(itself a striking and dramatic assertion, given its removal of humans from biblical chronologies), exactly when humans had first appeared in Europe, and how they had developed, was still a major controversy.[12] Dawkins positioned humans deep in the geological past, first entering the landscape in the Pleistocene. And here, archaeology blending with geology was crucial to understanding and defining ancient humanity.

Archaeology

With the study of human antiquity, methods moved to 'archaeology', and human artefacts and (if found) skeletal remains. Dawkins thought that early humans were locked in a grim and forbidding environment, and were made known through the practice of 'cave hunting' – a term used as the title of Dawkins's major popular work. A review of the book in *The Spectator* stated that 'sober science has nowhere more conspicuously displaced mystic fancy than in the popular ideas as to caverns. Their gloomy passages no longer lead the explorer bodily down to the depths of Hades, but they carry him mentally back into the far recesses of time.'[13] The *Liverpool Daily Albion* described how cave hunting 'is nearly as much an invention of the nineteenth century as photography, the electric telegraph, and the spectroscope', and allowed practitioners to 'combine keen excitement and hard climbing with results of by no means low scientific interest, and they can be enjoyed without the infliction of pain on any living animal'.[14]

A report in the *Daily News* described one 'cave hunt' in Denbighshire in idyllic terms, as a social event revealing mysteries of the human and animal past. Encouraging the reader to imagine 'a summer's day in Wales, and a party of ladies and gentlemen issuing forth on this scientific quest', they would see a 'small wood-covered hill rising abruptly out of the fields of yellowing corn', with 'three or four deep and dark caverns' where 'a group of labourers are to be seen hard at work with pickaxes and shovels'.

> But where is the sport? The 'game' is to be unearthed by the huntsmen from the heap of debris brought out every five minutes in baskets from the interior of the cavern ... the visitors fling themselves with scientific ardour – holding up now a wild-boar's tooth, and now a mysterious bit of polished stone, to the admiration and wonder of the company.[15]

'Admiration and wonder' were key aspects of this activity, as was an analysis of the objects being taken out of the caves by the unnamed labourers, which were discussed in ways that mirrored popular parlour games such as charades. While the deep past may have been monstrous and savage, the modern world, with its science and recreations, was picturesque and civilised. And the practice of cave hunting allowed a connection with the natural world that was free from modern animal suffering, although one that potentially transported the viewer back to darker and more savage times.

Despite this cosy tone, Dawkins's arguments around the ancient human past had darker currents. He used his studies of prehistoric artefacts and human skulls and bones to present a view of human development that was as layered and stratified as geological landscapes were, tied to the natural sciences and Victorian views of race and human inequality. Across his works, he posited that Europe had been inhabited by a series of human 'races', each tied to particular climates and animals, and linked with modern peoples judged as being at different levels of civilisation. First were the 'river drift hunters', equipped only with chipped flint tools and living alongside warm-adapted animals of an African type, such as elephants, rhinos and lions. They were judged – following Victorian colonial and racial ideologies – as akin to Aboriginal Australians (notably, on a round-the-world journey undertaken in 1875, Dawkins collected large numbers of racialised photos of Indigenous people when in Australia). Next came the 'cave men' who for Dawkins were not the modern stereotype of shambling ape-men, but were cold-adapted types, linked with the inhabitants of the modern Arctic. These had come to Britain in the Ice Age, alongside mammoths, musk oxen and marmots, and they depicted these animals in the highly realist artworks that had been found in southern France. These two 'lowest' human strata were marked out racially and culturally, and believed now to be extinct in Europe, and their analogues only persisted in places regarded by Dawkins as remote and uncivilised.

Moving into later periods, archaeology and history became enmeshed as newer human groups were expressly described in civilisational terms. A key role was played by the third migrants, the 'Iberians' of the Neolithic, depicted as a small, dark-haired people connected to the Basques as speakers of a non-Indo-European language. They were crucial for bringing agriculture and the first domestic animals, and therefore the foundational elements for civilisation. Importantly for Dawkins, while the river drift and cave men were now extinct, the Iberians still persisted in parts of Britain, especially in Wales, where 'the small swarthy Welshman of Denbighshire is in every respect, except dress and language, identical with the Basque inhabitant of the Western Pyrenees'.[16] He argued that Neolithic society was warlike and violent, but partly excused this tendency in cultural terms, saying that we nowadays 'have innumerable outlets for our activity, – we have the Press, we have political and religious excitement, we have games and amusements, and penny lectures, and many other things in our complex civilisation, to relieve the monotony of life which these people did not possess'.[17]

Finally, after the Iberians came two further racial types. These are 'the tall, round- or broad-headed Celt ... composing the van of the great Aryan army', arriving in Britain in the Bronze Age and setting up new societies over the Iberians (and bringing their own domestic animals and art forms); and finally, the Germanic peoples in the Anglo-Saxon period, who were responsible for establishing the societies of eastern England and the Scottish lowlands – but only over a much deeper substratum of peoples.[18] This was not just shown in humans and culture, but in animals, as 'the small Celtic short-horn' cattle were replaced by large white cattle, 'identical with those of Chillingham', which are 'distributed throughout every part of Britain conquered by the English, while the Celtic short-horn only survives in those parts in which the British had taken refuge'.[19] In this way, Dawkins used archaeology and natural science to construct an image of British history defined by ideas of racial stratification, in which violence and conquest were key drivers in the 'progress' of civilisation, and humans were persistently linked with animals. While the modern world might be genteel and full of amusements, older societies were violent, threatened and doomed.

History

Dawkins's ideas of civilisation linked the geological, biological, archaeological and historical, with all means of knowing the past bound into one single structure. These carried on into later periods, although Dawkins did not continue the historical story himself. A widely repeated anecdote reported that while at Oxford, Dawkins and his friend John Richard Green agreed to split research on the prehistoric and historic periods between them – and Green would indeed later become an authority on Anglo-Saxon England, and presented this in his Short History of the English People in a decidedly popular vein.[20]

In a more direct way, Dawkins presented his views in the national press (both scholarly and public), lecture tours and talks throughout Manchester and the north-west. He taught numerous courses dramatising his ideas on geology, human archaeology and evolution. The syllabuses of some of his extra-mural courses survive in Buxton Museum and Archives, and covered topics such as Ecology and Evolution, the Ancient History of the Earth and the Tertiary Period. These included reflective questions for each week, which encouraged students to recapitulate Dawkins's own theories. Students were asked to 'Describe the Flora of the Coal-Measures, and Explain the formation of a Seam of Coal', 'Describe to what extent the Neolithic and Bronze inhabitants of Britain are now represented in the existing population', and 'Discuss the evidence as to the relation of the Cave-man to the Eskimos.'[21]

Boyd Dawkins's most visible legacy in the modern city of Manchester is the Manchester Museum. The museum today is very much a pluralising institution, simultaneously connected with research in the University, wider communities in the city, and a varied and diverse view of nature and culture. Some of these aspects were present in the initial vision for the museum, while others have shifted a great deal – and the museum's overall history has been told by Samuel Alberti.[22] It was set up as a civic institution in a late Victorian mode. A new building was designed by the famous architect Alfred Waterhouse, who had designed the British Museum (Natural History) in London, and opened in 1888. According to Dawkins, this new museum was intended to follow the 'New Museum Idea' linking public education and research, and to break away from 'the old type of Museum, with its curiosities and other objects, intended to excite wonder, horror or disgust', which 'survived even in Manchester until the beginning of the [eighteen] 'seventies'.[23]

The initial layout followed Dawkins's vision of the links between different ways of knowing the world, from the geological to the historical, and the improving nature of education for the Manchester public. As the galleries expanded, Dawkins wrote that 'the scheme of classification is based upon the two great principles of time and evolution', starting with minerals, then the history of the Earth, then 'the history of life as revealed in the rocks', and finally natural history, moving through plants, animals and humans.[24] The galleries were intended to show the continuity between different branches of knowledge, with the floorplan arranged so 'it is the only Museum in Britain in which the continuity of Geology and history is clearly set forth. You walk straight from the Tertiary collection to those of Crete and Egypt, the one into the other!'[25] Pride of place was also given to particular important specimens, such as the great stump of *Lepidodendron*, a Carboniferous plant (excavated in Clayton in West Yorkshire in the 1880s, and purchased for the museum by William Crawford Williamson, Professor of Natural History at Owens College), and many Pleistocene fossils from Creswell Crags, donated by Dawkins and exhibited alongside modern Inuit artworks.

In Dawkins's view, the museum had been successful, and provided an example for other museums around the world. He wrote how it had 'made it possible to establish the Natural History departments in the University on a firm basis, and to educate graduates who are now scientific leaders at home and in the Colonies'; it was connected with local education, with 2,000 schoolchildren using the museum in their classes; and 'has further been a centre for various societies, and has contributed to widen the outlook of the public and particularly of the working man'.[26] In support of this project, Dawkins also frequently gave further public lectures in the museum, on such topics as 'The Great Extinct Reptiles', 'The Dawn of Life' and 'Diamonds, Sapphires, Carnelians, Onyxes'.[27]

However, while the museum's arrangement had been established to be 'sufficiently elastic to find a place for any new development that may arise in the future', it did not survive.[28] In 1927 Dawkins wrote a memo decrying the proposal to remove the Tertiary collections from between the zoological and ancient Egyptian collections. He wrote that this would destroy the interlinked aspects of the collection. It

> will diminish the teaching value of the Museum to the University in Zoology and Palaeontology and its attraction to the public. It will destroy the unique character of the Museum viz. – the proof of the

continuity of Geology with History ... It will also interfere with the arrangement of the section of Applied Geology, so much needed by Engineers and Architects.[29]

Very little now survives in the museum of Dawkins's developmental schema, and the subjects and fields that he saw as intrinsically interlinked were broken up. This reflects changes in the intellectual and institutional culture of the twentieth century, as disciplinary fields became somewhat more consolidated, but also the move of the museum itself away from the defining view of one 'museum master', and towards more plural and varied perspectives. This should not lead us to mourn the loss of a golden age of interdisciplinarity in late Victorian and Edwardian Manchester. There are very definite areas where we would find these projects troubling – the patronising didactic construction, the use of natural history to racialise and belittle many human cultures judged as 'primitive', the exploitation of fossil fuel resources, and the frequent drives to annex or dominate particular areas of the past. These trends – in all of their complexity – are important to keep in mind when we think about the growth of scholarship and scholarly institutions, and their wider social and cultural role.

Notes

1 Ian Hesketh, *A History of Big History* (Cambridge: Cambridge University Press, 2023).
2 William Boyd Dawkins, *Early Man in Britain and his Place in the Tertiary Period* (London: Macmillan, 1880), p. 1.
3 Ibid., pp. 1–2.
4 Howard Falcon-Lang, 'Marie Stopes, the discovery of pteridosperms and the origin of carboniferous coal balls', *Earth Sciences History*, 27:1 (2008), 85.
5 Geoffrey Tweedale, 'Geology and industrial consultancy: Sir William Boyd Dawkins (1837–1929) and the Kent coalfield', *British Journal for the History of Science*, 24:2 (1991), 435–51.
6 These excavations are discussed in detail in Mark John White, *William Boyd Dawkins and the Victorian Science of Cave Hunting: Three Men in a Cavern*, illustrated edition (Barnsley: Pen & Sword, 2017), pp. 102–87.
7 Tweedale, 'Geology and industrial consultancy'; Anthony S. Travis, 'Engineering and politics: the Channel Tunnel in the 1880s', *Technology and Culture*, 32:3 (1991), 461–97.

8 Naomi Yuval-Naeh, 'Cultivating the Carboniferous: coal as a botanical curiosity in Victorian culture', *Victorian Studies*, 61:3 (2019), 419–45, doi: 10.2979/victorianstudies.61.3.03.

9 William Boyd Dawkins, 'On the geography of the British Isles in the Carboniferous Period', in *Report of the Fifty-Seventh Meeting of the British Association for the Advancement of Science* (1888), pp. 684–5.

10 William Boyd Dawkins, 'On coal', in *Science Lectures for the People, Second Series* (Manchester: J. Heywood, 1870), p. 3.

11 Ibid., p. 16.

12 A. Bowdoin Van Riper, *Men Among the Mammoths: Victorian Science and the Discovery of Human Prehistory* (Chicago: University of Chicago Press, 1993); Donald K. Grayson, *The Establishment of Human Antiquity* (New York: Academic Press. 1983).

13 'Cave-hunting', *Spectator*, 12 December 1874, p. 1568.

14 'Cave exploration', *Liverpool Daily Albion*, 30 November 1874.

15 *Daily News*, 29 October 1874, p. 5.

16 William Boyd Dawkins, *Cave Hunting: Researches on the Evidence of Caves Respecting the Early Inhabitants of Europe* (London: Macmillan, 1874), p. 225.

17 William Boyd Dawkins, *Our Earliest Ancestors in Britain: A Lecture Delivered in the Public Hall, Collyhurst, Manchester, January 18th, 1879* (Manchester: J. Heywood, 1879), p. 101.

18 Dawkins, *Early Man*, p. 343.

19 Ibid., p. 492.

20 John Richard Green, *A Short History of the English People* (London: Macmillan, 1874).

21 Buxton Museum and Archives, William Boyd Dawkins Collection (henceforth Dawkins Collection), box 61.

22 Samuel Alberti, *Nature and Culture: Objects, Disciplines and the Manchester Museum* (Manchester: Manchester University Press, 2009).

23 William Boyd Dawkins, 'The organisation of museums and art galleries in Manchester', *Manchester Memoirs*, 57:3 (1917), 3.

24 Ibid., 4.

25 Dawkins Collection, Box 61, William Boyd Dawkins, Memorandum on the History of the Museum [1927].

26 Dawkins Collection, Box 20, Memorandum on Museums in Britain, n.d.

27 Dawkins Collection, Box 61, Manchester Museum, Short Addresses to the Public, every Sunday, Session 1907–8 [1907].

28 Dawkins, 'The organisation of museums', 4.

29 Dawkins Collection, Memorandum on Museums in Britain.

6

Ancoats and lab coats: Sheridan Delépine and municipal public health

Michael Worboys

The term 'Town and Gown' has its origins in tensions, sometimes riots, between students and the local population in the ancient universities of Edinburgh, Oxford and Cambridge. However, by the late nineteenth century, following the creation of so-called civic universities, 'Town and Gown' were cooperating, with higher education seen as contributing to the local economy and culture. Owens College became the Victoria University of Manchester in 1903 and was a model 'civic'. John Owens's first condition in his endowment for the college was that it should 'serve the needs of the city, county or region'.[1] The trustees expressed the same sentiment, wanting the college to 'possess the capability of providing what may be found to be the wants of the community'.[2] Initially, the focus was on teaching and training local men, and from 1883 women, in the knowledge and skills to work in the professions, industry, commerce and schools. Students organised to undertake good works in the community through the University Settlement movement, which in Manchester established its base in Ancoats.[3]

Research became important in the 1870s, first in the sciences, led by Henry Roscoe, and then across other academic departments. Staff and equipment became a resource for local industries for advice and problem solving. The arrangements were mostly ad hoc, but the Department of Pathology and Public Health under Sheridan Delépine was an exception.[4] From 1892 he developed a unique

6.1 Sheridan Delépine and the staff of the Public Health Laboratory, 1901. Courtesy of the University of Manchester.

relationship with agencies in the city and the region, creating a public health laboratory that provided routine investigative and diagnostic services and investigations of specific problems. It was a commercial operation, bringing prestige to the University. Such was its success that in 1905 it moved to a large Delépine-designed building situated between the new Manchester Royal Infirmary (MRI) and St Mary's Hospital off Oxford Road. When the laboratory opened in 1905 Lord Lister (the surgeon Joseph Lister) congratulated Delépine on his enterprise, saying that it would provide an immense service to Manchester and bring renown to the University.

The new laboratory embodied Delépine's vision of a reformed medicine based on prevention, not treatment. He represented and promoted new ideas, approaches and practices developed from work on germ theories of disease and laboratory medicine.[5] His focus was on infectious diseases, then the largest causes of morbidity and mortality. His vision was that the new ideas and tools of bacteriology could be used to make public health more effective. Instead of looking towards general clean-ups of insanitary environments, the actions of Medical Officers of Health (MOHs) could be more targeted and effective. Following laboratory diagnosis of specific germs, they could use laboratory-proven disinfectants and modes of sterilisation. Bacteriologists could give precise diagnoses of who should be isolated and for how long, and aid with the new antitoxin treatment for the childhood killer of diphtheria.

Bacteriology was a new venture for Delépine; he had previously worked in pathological anatomy – the study of diseased tissues. He was born in Switzerland and educated in Paris, Lausanne and Geneva before studying medicine in Edinburgh.[6] He graduated in 1882 and served as assistant to several of the leading pathologists of the day, including D. J. Hamilton, William Rutherford Sanders and Sir William Turner in Edinburgh, and then Sir Andrew Clark at St George's Hospital, London. Between 1882 and 1891 Delépine undertook much of Clark's routine pathological work and was appointed pathologist and lecturer in physiology and pathology at St George's. During this period, he established a reputation for his studies of the lungs, the liver and septic infections. The latter introduced him to bacteriology, a rapidly growing area of pathology, moving its focus from the results and processes of disease to their causes.

Cholera, milk and typhoid fever

Delépine joined Owens College early in 1892 and was appointed to the newly endowed Proctor Chair of Pathology and Morbid Anatomy. Manchester had gained a rising star. At his farewell dinner in London, he was toasted by leading medical scientists Victor Horsley, German Sims Woodhead and Lauder Brunton.[7] In 1892 there was a major cholera epidemic in Hamburg. Britain's east coast ports were seen to be vulnerable, as were Manchester and Liverpool, because of the numbers of people and the amount of goods that travelled by train across the Pennines. It was Delépine's first opportunity to show the value of the new public health.

Cholera was by then recognised as a water-borne bacterial infection; hence MOHs focused on water supplies and human carriers, who would be infection spreaders. Outbreaks of summer diarrhoea complicated the epidemiological picture, but bacteriology had the methods to differentiate the two diseases. Delépine undertook such tests for John Tatham, Manchester's MOH.[8] To much relief, he confirmed just three cases of cholera, confounding 'unauthorised and sensational reports' about its prevalence. He acted 'voluntarily and without previous agreement' and was thanked by the city for his 'courtesy and public-spiritedness'.[9] An editorial in the *Lancet* argued that this work added support to the already planned extension of pathology laboratories and 'hoped that the Corporation through their representatives on the governing body of the College will take care that the department does not lack the necessary funds for its satisfactory equipment with the most recent and best apparatus for bacteriological study and research'.[10]

At the end of the year Delépine gave a talk to the Manchester Medical Society on the bacteriological diagnosis of cholera.[11] He stressed the difficulties of differentiating it from other diarrhoeal diseases by clinical and anatomical signs and the superiority of laboratory methods. However, he did not use standard microscopy or plate-culturing techniques but rather chemistry (the production of nitrites and indol in water) and inoculation in guinea pigs. He ended by reviewing the current state of knowledge of cholera, arguing that bacteriology had brought new questions as well as settling old ones, and, perhaps surprisingly, warning against any 'dogmatic assertion of the infallibility of the bacteriological diagnosis of Asiatic cholera'.

In 1894 James Niven, who had a particular interest in tuberculosis and the milk supply, was appointed the new MOH for the city.[12] He found a ready ally in Delépine, and they became good friends and colleagues. Niven had previously been MOH for Oldham and had a particular interest in the transmission of tuberculosis to humans from cattle via meat and milk.[13] Milk as a source of infection had long been a public health issue, principally its contamination by dirt causing diarrhoea, which caused high mortality in children.[14]

But there was a new threat, cattle with tuberculosis producing infected meat and milk.[15] Campaigning by MOHs and public fears led the government to appoint two Royal Commissions to inquire into the problem.[16]

While the Commissions deliberated, Niven instigated measures to improve Manchester's milk supply, relying on Delépine's laboratory to test samples from farms, dairies and shops.[17] The number of tests undertaken multiplied, and news of the service spread, leading MOHs across the north-west to send samples for testing. Such was the laboratory's profile that Lord Vernon asked Delépine to advise on eradicating the disease from the cattle on his Sudbury Hall estate in Shropshire. In September 1898 Delépine gave a talk at Vernon's estate to landowners and farmers.[18] He reviewed the new understanding of tuberculosis and how to control, if not eradicate it. He cited the cumulative results from testing on the incidence of *Tubercle bacilli* milk from urban dairies in Manchester and surrounding districts.

> The milk of at least 1 out of every 4 cows much affected with tuberculosis is capable of communicating tuberculosis ... The milk coming from country farms which are not under the constant supervision of the health authorities is capable of communicating tuberculosis in 17.6 per cent of the cases examined. A little more than 1 sample out of every 6 examined proved infectious.[19]

Manchester was not yet to the fore. Liverpool had shown what could be achieved by effective sanitary inspection, as only 6 per cent of dairies there were infected.

A significant problem facing MOHs was the milk supply from surrounding rural areas where they had no jurisdiction. Central government stalled on adopting the second Royal Commission's recommendations on inspections, leaving control to local authorities. Manchester now led the way, its measures becoming a model for other cities. With Delépine's support, Niven persuaded the city to pass an amendment to the Manchester Corporation (General Powers) Act. Parliamentary approval was required for what became known as the 'Manchester Milk Clauses'. The problem of jurisdiction was overcome by putting the responsibility on sellers. Officials were able 'to prosecute [and fine] anyone who knowingly sold milk from cows with tuberculosis of the udder'. Dairies supplying the city had to isolate infected cows and notify any cow with signs of tuberculosis of the udder. The integration of Delépine's laboratory into the scheme was clear, as inspectors had powers to send samples of milk or udder tissue for laboratory testing. These measures were copied in Liverpool, Birmingham, Leeds, Sheffield and Sunderland. By 1910 model milk clauses had been adopted by 102 local authorities, but in most areas they were unenforced.[20]

The number of milk inspections and tests undertaken in the Pathology Department grew.[21] In 1897–98 110 specimens were tested; in 1906 the number was 707. By 1908 over 5,000 Manchester samples had been tested, plus 1,500 from other local authorities from 2,079 farms. Positive tests were followed up, with farmers being advised on sanitary measures or prohibited from selling their milk. The incidence of disease in samples fell over the decade from 1897–98 from 16.1 per cent to 6.5 per cent.[22] Such was the success of the Manchester Milk Clauses that the Local Government Board asked Delépine to report on the city's experience so that local authorities across the country might learn from and emulate the measures. The key to success was the combination of skilful work by veterinary inspectors in Niven's department based in the Town Hall and the bacteriological examinations undertaken at the University.[23] One limitation of the Clauses was that farmers whose cows were found to be tuberculous were sending their milk elsewhere. Hence, a national Manchester-type scheme was needed.

Testing milk was only part of the laboratory's work. The Report for 1904 showed that 5,840 reports had been made: 3,135 for Manchester and 2,994 for 62 other authorities, with most made for Salford, Withington and Blackpool.[24] Just 11 per cent were for tuberculous milk. The largest number of tests were for diphtheria – 2,383, with 31 per cent positive. The disease mainly affected children, with positive cases sent to isolation hospitals.[25] Typhoid (enteric) fever cases came next with 1,407 tests, of which 41 per cent were positive. Affected adults and children were also isolated, and their water supply tested and homes disinfected. Domestic hygiene was a subject that Delépine had promoted since his appointment, first speaking on the subject to the Ancoats Healthy Homes Society in 1894.[26] The third largest group of tests was for the tubercle bacillus in sputum (1,111 with 34 per cent positive), with samples from the Corporation and the newly founded open-air sanatoria. The work was all funded on a fee-per-service basis. It was hard to say if fees covered costs, as Delépine and the University bursar kept separate accounts. The laboratory staff also undertook some research. Delépine found that routine work raised new questions and he was commissioned to make special investigations. For example, in 1904 Manchester Corporation paid for reports on bacteria in preserved foods, micro-organisms in sewer air and flies as disease carriers. Delépine's subject and geographical range were impressive: chemical analyses for arsenic in the Salford beer scare, veterinary diseases with anthrax in Cheshire, and foodstuffs with poisonous pork pies in Derby and Uttoxeter.[27]

The Public Health Laboratory

In 1895 the terms of Delépine's work for local sanitary authorities were that it should 'not interfere with his teaching and did not cause any expense to the College'.[28] Formally, it was 'a private concern', in which differences between income and expenditure had to be made up by Delépine, which he did from his own resources and consultancy fees. In 1902 the arrangement changed when the laboratory was made a university department, receiving £500 each year from the annual £4,000 grant that the Director of Higher Education at Manchester Corporation paid the University.[29]

Delépine maintained that his laboratory subsidised Medical School teaching as student fees did not cover costs. It taught pathology and bacteriology and put on postgraduate courses, many of which were taken part-time or on secondment by local MOHs.[30] The fourteen courses offered were 'practical', designed for public health staff and public analysts. In 1904 a diploma in Veterinary State Medicine was started to meet the local and regional demands for inspectors to apply the milk clauses. For the session 1904–05, there were 97 students, 68 of whom were new entries, taught in Manchester and Burnley. Teaching was demanding, as all courses involved laboratory work that required the preparation of equipment, specimens and reagents, as well as lectures, supervision and assessment.

At this time, testing was undertaken by six staff, while teaching drew on staff from departments across the University, including William H. Perkin for chemistry and Arthur Schuster for physics.[31] The mainstays of the public health teaching were Delépine and E. J. Sidebotham, who had joined in 1894 and was unpaid – he had a private income.

In 1904 the University agreed to build a new Delépine-designed building and grant a new status for his venture, the Public Health Laboratory (PHL). Two years previously, the department had moved from the Medical School down Oxford Road, to a building in Stanley Grove. This building was on land chosen for the MRI when it moved from Piccadilly Gardens, so another location was required. Delépine found a site further along Oxford Road in York Place, placing the laboratory between the MRI and St Mary's Hospital but away from the University. The foundation stone was laid in April 1904 and the PHL opened in October.[32] From a financial perspective, it was a speculative venture. The laboratory had built up a deficit because its fees-per-service were set low to match what sanitary authorities would pay. The new building costs were covered by a payment from the MRI for the Stanley Grove site (£5,855) and a donation from the engineer W. J. Crossley (£2,000), but the accumulated debt of £5,270 remained.[33] The University Council was satisfied with the PHL's financial viability. While keen to reduce debt, it asked for and was given an exemption from the rates, and was prepared to carry its debt because of the profile and value of its work in the city.

Delépine's design was impressive and state-of-the-art. The PHL had a 50-yard frontage and was set in just over an acre of land.[34] There were two floors, with a central corridor end to end.[35] The laboratories were on the cooler north side: clinical pathology, bacteriology, research, municipal investigation and preparation, with offices and teaching rooms on the southern side. All the rooms had curved corners to avoid dust accumulation and enable efficient

6.2 The Public Health Laboratory, York Place, Manchester. Courtesy of the University of Manchester.

disinfection. In addition to architecture, Delépine had again used his engineering skills, designing medical devices: a gas-sphygmoscope that measured the pulse, and a microtome for slicing tissues for microscopy.[36]

The opening ceremony for the PHL took place on the morning of 27 January 1905 and was a grand civic occasion.[37] In the afternoon, honorary degrees were awarded to 'Men of World's Fame': the French microbiologist Albert Calmette and the explorer Robert Scott.[38] The great and the good of British medicine and medical science sent their good wishes, as did other distinguished scientists from abroad. The mayor of Manchester spoke of the valuable work the laboratory had done for the city and the security it provided against future epidemics. He added, perhaps prompted, that 'it would be an advantage in the practical prevention of infection if Professor Delépine were free to devote more time and money to research both personal and in the direction of others'. Pleas were also made for funds to clear the laboratory's debt.[39] Local councils were to be asked to pay higher fees and make donations, but in the meantime, the University was prepared to underwrite its flagship municipal venture.

In the new laboratory, the routine work for local authorities continued, increasing from 7,480 tests in 1905–06 to 14,551 in 1913–14.[40] The absence of local epidemics and improvements in milk safety lowered its public profile. No endowment for research was forthcoming, but Delépine's work changed. His reputation enabled him to become a bacteriological and pathological consultant on various health issues and increasingly on the national stage. He was in demand at every level, from government public health policies to serving as an expert witness in a coroner's court in Wigan.[41] He continued to advise on the dangers of tuberculous milk from cows and gave evidence to the Departmental Committee on Tuberculosis in 1913.[42] The committee mainly considered treatment in open-air sanatoria, but Delépine offered a different approach based on his pathological and bacteriological research, favouring prevention rather than treatment. First, post mortems had shown that around 90 per cent of adults over 30 years of age had tubercular lesions, though in most people they were arrested and did not lead to disease. Sanatoria were said to produce cures with incipient and early cases, and this promise had attracted local authority and philanthropic support. However, Delépine argued that the high prevalence of early and hidden infection revealed by post mortems meant that sanatorium treatment was impractical and unaffordable. The experience with milk had shown the value of preventing infection, so a better use of sanatoria would be to isolate the most infectious, advanced cases. He conceded, however, that this was unlikely to happen. Funding had been attracted to sanatoria because they offered the hope of a cure; there would be much less support if they became seen as homes for incurables.

Work on the contamination of water supplies grew and included chemical investigations, for example, poisoning from lead piping.[43] Food contamination was a continuing demand, with analyses of ice cream, meat, mussels and oysters. The latter took Delépine to Poole Harbour on the south coast.[44] However, most work was undertaken for the city and for local authorities in Lancashire.

With the outbreak of the First World War, Delépine put his expertise at the service of the nation on a typically wide range of problems.[45] In 1914 he was called upon by the army to investigate an outbreak of typhoid in soldiers inoculated against the disease and on methods of sterilising wound dressings on a large scale. Later, he reported on the pathological effects of gas warfare agents Arsine, Lewisite and Phosgene, and on diseases suffered by soldiers at the front: trench foot, frostbite and ringworm.[46] His knowledge of arsenic poisoning was used to monitor the side effects of Salvarsan treatment for syphilis, and he advised on putting together venereal disease diagnosis and treatment kits for use at the Front. Once again showing his engineering expertise, he invented a portable refrigerator and incubator boxes for use in laboratories at the Front.[47]

The number of routine tests did not decline during the war and continued to rise again afterwards. In 1920–21, tests for diphtheria (9,716, 16 per cent positive) and tuberculous milk (1,126, 11 per cent positive) continued to be in demand, with a growing call for reports on tuberculous sputum (5,611, 28 per cent positive). One hundred and fifteen local authorities and seven hospitals used the PHL's services. It had an income of £7,364, and in most years it made a small surplus that enabled it to clear its debt. The work was done with a staff of six scientists and thirteen others. However, all was not well. Delépine's health was failing, and he was often absent. Change was coming. The University Council wanted the laboratory to make more money, perhaps prompted by testing in the Pathology Department, which made 100 per cent profit on the venereal disease tests it was doing for the army and the Ministry of Health.

Delépine responded by producing an audit showing that the PHL had only cost the University £3,880 over 1892–1920, a small price for the good it had done in the city and the region and for the standing of the University.[48] Still worried, he wrote to Niven suggesting that the city buy the laboratory and take it in-house. The University Council was surprised and annoyed at the suggestion.[49] It wanted to continue support, though with regular payments rather than unpredictable fee-per-service income. Delépine died in November 1921 while the future of the PHL was undecided.

The next head of department, William Topley, was unhappy with the department's public health and commercial orientation. He changed its name to Bacteriology and Prevention Medicine, separating the PHL's testing work from its teaching and research. This arrangement persisted until the PHL closed in 1948 when its work was absorbed into the new Public Health Laboratory Service (PHLS). Ironically, the PHLS developed from the Emergency Public Health Laboratory Service set up by Graham Wilson and Topley in 1939.[50]

Conclusion

Delépine's obituaries all stressed his work for Manchester and the surrounding region. A University Senate resolution stated that his 'unstinted labours ... to the organisation and development of the Public Health Department ... had confirmed lasting benefit upon the community'.[51] The *Manchester Guardian* observed that 'Professor Delépine rendered the most distinguished service in all that relates to the hygiene and well-being of the community in and around Manchester.'[52] The *Lancet* noted how staff and students in his laboratory 'connected with the immediate sanitary needs of Manchester, Salford, and the surrounding districts'.[53] A commemorative bust was unveiled in the Medical School by Sir George Newman, the chief medical officer of the Ministry of Health, in December 1923.[54] In his eulogy, he stressed the key idea that directed Delépine's scientific work.

> An enthusiastic believer in preventive medicine and its potential influence in the life of human society, he became an inspirer and encourager of others. Though he failed to win some of the world's glittering prizes, he became something of a national leader in his subject, devoting his life and sacrificing much to establishing a great school of public health in that university.[55]

He also pointed to his commitment to public service, noting the collaboration between Delépine and Niven: 'a partnership of scientific and public work which had been of inestimable value to Manchester and the whole country'. The PHL exemplified how a civic university could work with city agencies for mutual benefit: diseases were prevented and public health improved, alongside the development of a state-of-the-art scientific institution.

A decade earlier, an editorial in the *Manchester Courier*, titled 'The Business University', had commented on the place of the University in the city's economy. The PHL was singled out for special praise.

> [T]here is no department of the university to which the public is so much indebted for practical benefit, and for the provision of practical workers on behalf of public health, than to the Public Health Department of the University and its extensive laboratories in York Place, under the directorship of Professor Delépine. Here doctors, both of man and animal come to take special courses to fit them for public positions under the State and Public Authorities. Beyond this there is hardly an Urban or Rural Sanitary Authority in Lancashire or Cheshire – beside many in more distant places – that has not sought the Department's advice. In doing so, too, they have been training men for Public Health work, School Hygiene, Factory Hygiene, Veterinary State Medicine, and so on, while research and postgraduate work is carried on whose results may be of incalculable benefit to humanity.[56]

The PHL graduate was praised for being 'scientifically practical', not 'an unpractical "University man"'.[57] In other words, the city needed graduates in lab coats, not gowns.

Notes

1 H. B. Charlton, *Portrait of a University, 1851–1951* (Manchester: Manchester University Press, 1951), p. 45.
2 Ibid., p. 52.
3 M. E. Rose, 'Settlement of university men in great towns: University Settlements in Manchester and Liverpool', *Transactions of the Historic Society of Lancashire and Cheshire*, 139 (1990), 137–60.
4 W. J. Elwood and A. F. Tuxford, *Some Manchester Doctors: A Biographical Collection to Mark the 150th Anniversary of the Manchester Medical Society* (Manchester: Manchester University Press, 1984), pp. 107–12.

5 Michael Worboys, *Spreading Germs: Disease Theories and Medical Practice in Britain 1865–1900* (Cambridge: Cambridge University Press, 2000), pp. 234–76.

6 Michael Worboys, 'Delépine, Auguste Sheridan (1855–1921), pathologist and bacteriologist', *ODNB*, doi: 10.1093/ref:odnb/57113; G. Sims Woodhead, 'Sheridan Delépine. Born 1st January 1855–Died 13th November 1921', *Journal of Pathology*, 25 (1922), 113–17.

7 'Farewell dinner to Professor Sheridan Delépine', *Lancet*, 139:3570 (30 January 1892), 263.

8 John Tatham, 'Cholera in Manchester', *Lancet*, 142:3657 (30 September 1893), 813.

9 'Bacteriology at Owens College', *Lancet*, 142:3661 (28 October 1893), 1090–1.

10 Ibid., 1091.

11 Sheridan Delépine, 'The bacteriological diagnosis of cholera', *Lancet*, 142:3669 (23 December 1893), 1572.

12 Joan Mottram, 'Niven, James (1851–1925), public health administrator', *ODNB*, doi: 10.1093/ref:odnb/57128.

13 'Manchester Medico-Ethical Association', *Manchester Times*, 15 April 1992, p. 2.

14 P. J. Atkins, 'White poison? The consequences of milk consumption, 1850–1930', *Social History of Medicine*, 5 (1993), 207–27; Jacob Steere-Williams, *The Filth Disease: Typhoid Fever and the Practices of Epidemiology in Victorian England* (Rochester, NY: University of Rochester Press, 2020).

15 P. J. Atkins, 'Milk consumption and tuberculosis in Britain, 1850–1950', in Alexander Fenton (ed.), *Order and Disorder: The Health Implications of Eating and Drinking in the Nineteenth and Twentieth Centuries* (East Linton: Tuckwell Press, 2000), pp. 83–95.

16 Royal Commission to Inquire into Effect of Food Derived from Tuberculous Animals on Human Health, PP 1895 [C.7703] XXXV: 615, and PP 1896 [C.7992] XLVI: 11; Royal Commission to Inquire into Administrative Procedures for Controlling Danger to Man Through Use as Food of Meat and Milk of Tuberculous Animals, PP 1898 [C.8824] XLIX:333 and [C.8831] XLIX: 365.

17 James Niven, *On the Improvement of Milk Supply in Manchester* (Manchester: Heywood, 1896).

18 Sheridan Delépine, 'A lecture on tuberculosis and the milk supply, with some general remarks on the dangers of bad milk', *Lancet*, 152:3917 (17 September 1898), 733–8.

19 Ibid., 737.

20 W. G. Savage, *Milk and the Public Health* (London: Macmillan, 1912), pp. 330–1.

21 Sheridan Delépine, 'The Manchester milk supply from a public health perspective', *Transactions of the Manchester Statistical Society*, 56 (1909–10), 1–2.

22 Thirty-eighth Annual Report of the Local Government Board, 1908–09. Supplement containing the report of the medical officer for 1908–9. PP 1909 [Cd.4935], xxviii, 877.

23 Ibid., p. 879.
24 Sheridan Delépine (ed.), *Archives of the Public Health Laboratory of the Victoria University of Manchester*, vol. 1 (Manchester: Manchester University Press, 1906), pp. 401–6.
25 J. V. Pickstone, *Medicine and Industrial Society: A History of Hospital Development in Manchester and its Region, 1752–1946* (Manchester: Manchester University Press, 1985), pp. 173–81.
26 'Ancoats Healthy Homes Society', *Manchester Guardian*, 15 December 1894, p. 4. See also G. Mooney, *Intrusive Interventions: Public Health, Domestic Space, and Infectious Disease Surveillance in England, 1840–1914* (Rochester, NY: University of Rochester Press, 2015), pp. 144–7.
27 'Poisoning extraordinary', *Manchester Times*, 30 November 1900; Matthew Copping, 'Death in the beer-glass: the Manchester arsenic-in-beer epidemic of 1900–01 and the long-term poisoning of beer', *Journal of Brewing History*, 132 (2009), 31–57; 'Anthrax in Cheshire', *Manchester Courier*, 29 September 1903, p 8; 'Derby poisoning sensation', *Derby Daily Telegraph*, 15 September 1902; 'Professor Delépine's report', *Derby Daily Telegraph*, 16 September 1902. Delépine christened the pork-pie bacillus *Enteriditis derbiiensis*, 'a rare specimen … never been met with before or since', *Derby Daily Telegraph*, 13 December 1923, p. 4.
28 University of Manchester Archives (UML), SPH/1/1/1, University of Manchester, Public Health Laboratory, Income and Expenditure, 1892–1920, p. 2.
29 UML, UCO/1/3, Minutes of Council Meeting, 27 April 1904, p. 154.
30 Delépine (ed.), *Archives*, pp. 409–14.
31 Ibid., p. 398.
32 'Manchester Public Health Laboratories', *Manchester Guardian*, 25 January 1905, p. 5.
33 William Crossley was chair of the PHL's advisory committee. With his brother, he founded Crossley Brothers, a major engineering employer in the city and supporter of the Ship Canal. He was the Liberal Party MP for Altrincham between 1906 and 1910. His philanthropic work included serving as chairman of the Manchester Hospital for Consumption and Diseases of the Throat and Chest and giving the funds to build the Delamere Forest Sanatorium. See K. A. Barlow, 'Crossley, Francis William (1839–1897), engine manufacturer and philanthropist', *ODNB*, doi: 10.1093/ref:odnb/46858.
34 'Manchester Public Health Laboratory to be opened on Friday', *Manchester Guardian*, 25 January 1905, p. 5.
35 W. J. Crossley, 'Building of the new University Laboratory (York Place)', in Delépine (ed.), *Archives*, pp. 394–7.
36 'Professor Delépine', *Manchester Guardian*, 14 November 1921, p. 12.
37 'Safeguarding public-health: bacteriological research work in Manchester: new laboratories', *Manchester Guardian*, 28 January 1905, p. 6.
38 'Victoria University: the new Pathological Laboratories', *Manchester Courier*, 28 January 1905, p. 9.
39 'Manchester Public Health Laboratory to be opened on Friday', *Manchester Guardian*, 25 January 1905, p. 5.

40 UML, SPH/1/2/3, Annual Report of the Public Health Laboratory, 1920–21, p. 590.
41 'The wrong bottle', *Evening Telegraph*, 4 February 1910, p. 4.
42 Memorandum submitted by Sheridan Delépine, Final Report of the Departmental Committee on Tuberculosis: Volume II, Appendix, (1913), PP, 1912–13, (Cd.6654), xlviii, pp. 72–7.
43 'Water from the hills', *Manchester Courier*, 2 May 1908, p. 8; Sheridan Delépine et al., *Reports on the Investigation of Cases of Lead Poisoning* (London: Home Office, 1913).
44 'Poole oyster fishery', *Western Gazette*, 27 December 1907, p. 10; 'The Poole oyster', *Western Gazette*, 19 September 1919, p. 8.
45 A list of Delépine's reports is given in R. M. Stirland, Auguste Sheridan Delépine, 1855–1921, Personal and Family Documents, pp. 96–7. Available at UML, SPH/2/2.1.
46 Sheridan Delépine, 'Remarks on some of the effects of exposure to wet cold and their prevention', *British Medical Journal*, 2:2868 (18 December 1915), 888–92, doi: 10.1136/bmj.2.2868.888.
47 Sheridan Delépine, 'A simple portable refrigerating box and its use as a temporary incubator in military work', *British Medical Journal*, 1:2890 (20 May 1916), 718, doi: 10.1136/bmj.1.2890.718.
48 UML, SPH/1/1/1, Public Health Laboratory, Income and Expenditure, 1892–1920, pp. 2–16.
49 UML, UCO/1/11, Minutes of the University Council, 27 July 1921, p. 236; 7 December 1921, p. 282.
50 G. S. Wilson, 'The Public Health Laboratory Service: origin and development of public health laboratories', *British Medical Journal*, 1:4553 (10 April 1948), 677–82.
51 'Sheridan Delépine, M.Sc., M.B.', *British Medical Journal*, 2:3178 (26 November 1921), 921, doi: 10.1136/bmj.2.3178.921.
52 'Professor Delépine', *Manchester Guardian*, 14 November 1921, p. 12.
53 'Obituary: Sheridan Delépine', *Lancet*, 198:5125 (19 November 1921), 1080.
54 'A pioneer of public health', *Manchester Guardian*, 13 December 1923, p. 11.
55 Ibid.
56 'The business university', *Manchester Courier*, 29 September 1913, p. 6.
57 Ibid.

7

'In the grey-built city of the mind / Wave the green boughs of a few hostage powers': Eva Gore-Booth, Esther Roper and the Manchester University Settlement in Ancoats, 1896–1907

John McAuliffe

In their essay collection *Culture in Manchester: Institutions and Urban Change since 1850*, Mike Savage and Janet Wolff attribute the 'continued and varied cultural vibrancy of the city' to the fact that

> alongside the Hallé Orchestra, the Literary and Philosophical Society, the City Art Gallery and the Manchester Museum there have thrived smaller musical societies and theatre groups, as well as local popular cultural practices and pursuits [...] Middle-class culture is always somehow in dialogue and exchange with working-class culture. This relationship is especially striking in Manchester.[1]

At the turn of the last century, Owens College developed an innovative civic platform to enable just such dialogue and exchange, supporting new connections between University lecturers or graduates and working-class communities in Ancoats. The University's social experiment, the University Settlement, inspired by John Ruskin's critical vision of political economy, would last for decades. The Settlement shaped new ideas and was shaped in turn by how this platform was used, especially in its first decade. Among its Associates were two women, Esther Roper and the poet Eva Gore-Booth, whose work in Manchester is only recently re-emerging as a significant achievement.

Ruskin in Manchester

On 24 March 1904 the *Manchester Guardian* reported on 'The Ruskin Exhibition', which had just opened at the City Art Gallery. Gathering examples of Ruskin's own work in literature and visual art, the exhibition connected them to the work of those who had influenced Ruskin and to younger artists who had been influenced by him. Manchester's civic institutions were not averse to a public and celebratory occasion, and the many speakers included the Lord Mayors of Manchester and Salford, the bishop of Salford, the philanthropist Thomas Horsfall, and the vice-chancellor of the University of Manchester, Dr Alfred Hopkinson, each of whom remembered Ruskin's relationship with the city.[2] Hopkinson found it 'peculiarly fitting that Manchester should take an active part in a demonstration in honour of one of the greatest teachers England had produced, for it was in Manchester that some of the most characteristic teachings of Ruskin were first heard'.[3] Ruskin had indeed had a long, often antagonistic and productive relationship with Manchester: he delivered two lectures on 'The Political Economy of Art' at the great 'Art Treasures Exhibition' of 1857, returning in 1859 to deliver a lecture on 'The Unity of Art' at the School of Art, and in 1864 to deliver two lectures at Rusholme Public Hall, 'On King's Treasuries' and 'Queen's Gardens', which would be published together as his best-selling book *Sesame and Lilies* in 1865. Ruskin's critical importance to the city's image of itself had led to the inaugural Ruskin Society being established in the city in 1879, and in 1886, as a response to Ruskin's directive to see the arts as an intrinsic good and benefit to individuals and communities, the establishment of the Ancoats Art Museum in Ancoats Hall. As the *Guardian* reporter noted of the vice-chancellor:

> When asked to take part in the opening it seemed appropriate that the representative of our University should do what was possible to give expression to the sympathy with which those who took part in higher education must feel for such an effort as this exhibition to make known to our citizens and to excite more interest in the work of a great teacher [...] to Ruskin we owed an enormous debt for his social, political and moral teaching.[4]

Ruskin, in other words, was a foundational influence for the men who worked at Owens College and for those who were responsible for the city's educational initiatives, including the development of the Victoria University of Manchester.

Alongside Hopkinson, the University was represented at the Town Hall lunch by a number of his colleagues with a variety of links to the Art Gallery committee, including Horsfall and Ernest Phythian, who had, in the previous decade, been key figures in the University's work in Ancoats, still one of the city's poorest areas at that time. This work had initially involved the short-lived development of a house for University staff and city residents, formally called Ruskin House, and aligned to the Ancoats Art Museum. Soon after this they were involved in the initial establishment of the University Settlement, and then the stewardship of its home for decades afterwards, the Round House on Every Street. The University Settlement idea had originated in Toynbee Hall, which accommodated middle-class workers and university lecturers in the deprived working-class East End of London, with a view to knowledge exchange and social reform. At the Manchester Settlement, likewise, the warden and residents were required, as an earlier *Manchester Guardian* report put it, to do 'some social or educational work themselves' in the Ancoats area.[5] This report further defined the University Settlement's mission as to ensure that

> working men are not to be educated out of their class, but in it and of it. The end they are to have in view is the betterment of those who are less capable or less fortunate than themselves. What they have received they are to communicate. They are to be social missionaries, messengers from the living to the dead, agents in behalf of the re-union and co-operation of society [...] and they have the helpful guidance and companionship of the warden, the members of the University Settlement, and many of the Professors at the Owens College.[6]

If Ruskin, also a major influence on Toynbee, was a key figure for those who envisioned this civic role for the University in the city, the Settlement was also crucially driven by another set of ideas that were reshaping the University in the late nineteenth century, ideas that led to the University teaching and graduating women students, and becoming, through its debating societies, a platform for those campaigning for women's suffrage.

In fact the genesis of the University Settlement in Ancoats came from a debate at Owens College, where the principal, Hopkinson's predecessor Dr Adolphus Ward, suggested that he and his colleagues had a responsibility to more than their research and teaching, an idea to which the visiting warden of Toynbee Hall, Canon Barnett, responded with an outline of how he had developed Oxford University's Toynbee Hall Settlement in Whitechapel. At the meeting, alumni, staff and current students were voted on to the executive committee, including Horsfall, Phythian and notable academics including Professor Samuel Alexander and Professor Thomas Tout, who would go on to deliver evening lectures at the Settlement for decades afterwards.

In his study of Victorian Manchester, Michael E. Rose discusses the University Settlement especially in relation to the role that women played in its activities, pointing out that University graduates who served as part of the University's Court (its supreme governing body) were also key figures for the Settlement, including Mary Tout (Thomas Tout's wife) and the first joint warden, Alice Crompton, herself a niece of an early campaigner for promoting education and a member of the city's first Schools Board, Lydia Becker. In line with this proactive approach to women's representation in governance positions, it was also agreed that the Settlement committee should have an equal number of women as representatives, and among the six women selected was one of the first to receive a degree in Manchester, Esther Roper.

The situation of the Settlement followed the same rationale as Toynbee Hall in Whitechapel. Ancoats' reputation as a bleak, industrial suburb reached back as least as far as Engels' 1849 description of it as 'Hell upon Earth', and the University Settlement was not the first intervention there; however, it differed from other 'missions' in that, as was the case with Owens College itself,[7] it was not primarily driven by Christian or evangelising aims.[8] The Ancoats Brotherhood ran a regular series of talks at New Islington

Hall, with illustrious speakers including William Morris, and Horsfall's Ancoats Art Museum was already a fixed point of reference, while there also existed an Ancoats Recreation Movement and the Mill Girls Institute. Horsfall's involvement in the Settlement, and his volunteering of accommodation, was also decisive in the speedy choice of location. But his Ancoats Art Museum was not in itself enough to support the plan. The Settlement's first annual report, in 1897, includes a telling list of those who contributed to the Settlement's costs, a familiar roll call which included Horsfall, but also others of the city's great and good: Rylands, Worthington, Haworth, Ashton. In 1900 further substantial donations, from Mrs Rylands and Mrs Worthington, enabled the purchase of independent quarters, near the museum, at the Round House on Every Street. Horsfall, however, remained vital to its development. His museum would enter into a Memorandum of Agreement with the University Settlement in 1897, as recorded in the Settlement's first annual report, which is clear about the joint purpose and vision of the new organisation, an arrangement that was further strengthened by the legal amalgamation of the two organisations in 1902. The University's connection to the city's 'first suburb' was securely in place.

An educational mission

The Settlement's distinctiveness lay in its connection to an institution, Owens College, whose focus was on research and teaching, and which was constitutionally non-religious. In a study of Horsfall's various civic and town planning projects in Manchester, Michael Harrison observes how this chimed also with the museum:

> To look at the circle around T.C. Horsfall inevitably produces its own bias. It gives emphasis to 'preventive work and education' rather than rescue work. It stresses public health efforts rather than the manifold medical charities that existed in Manchester. It gives greater weight to cultural and recreational schemes rather than those of a more evangelical nature.[9]

While the first historian of the Settlement, Mary Danvers Stocks, was cautious about identifying the Toynbee Hall Settlement as the

7.1 The Round House, Ancoats, home of the Manchester University Settlement. Courtesy of the University of Manchester.

progenitor of Manchester's Ancoats Settlement, she was clear that both shared their origins in moving away from a Christian mission to modelling education as the route to equipping poorer communities with more opportunities. The Settlement's constitution and reports on activities demonstrate the interplay of different theories of higher education, alongside the emphasis on socially responsible civic values. The evolution of social science and advocacy at the Settlement was not, of course, entirely in tune with the Settlement's varied activities, although the tensions undoubtedly generated its organisational momentum, and framed the concerns of its better-known wardens and associates. Stocks's account shows how the Settlement's initiatives responded to the competing contemporary arguments about social reform played out, as she put it, between Beatrice and Sidney Webb's work on co-operative societies and central planning as opposed to the individual-focused self-help advocated by Octavia Hill:

In its cultivation of aesthetic values, in its dogged pursuit by road and rail of Nature's wonders, in its preoccupation with Italian art and literature, [the University Settlement] reflected the ideals of Octavia Hill, disciple of Ruskin and parent of the National Trust. But in its growing concern for the development of social services and its obvious turn to the left in politics, it was incontestably Webbian.[10]

The Settlement, as many of those who have written about it attest, was a contact zone, a place where new ideas and terms rubbed up against one another, at a time when change with regard to the role of the state in relation to education and welfare, and suffrage, was on the horizon.

The new constitution which was written to effect the amalgamation of the museum and Settlement in 1902 indicates the multipronged approach of the Settlement to questions of education and knowledge exchange and of inter-class networks. Its ambitious commitment to various activities maintained Ruskin's positive faith in culture as a benefit for the individual and for civic society, while also developing the Settlement's role as an outpost for the development of social sciences and social science research. The legal merger named as three of its objects:

> To disseminate and nourish a healthy love of Nature and of the best in Art, Music, Literature, and Science amongst the industrial population of Manchester, and especially of Ancoats, by the exhibition at the Art Museum and elsewhere, and by the loan to schools and other institutions, of pictures, casts, and other works of Art, and by the organization of lectures, classes, concerts, boys' and girls' clubs, debating and literary societies, flower shows, country rambles and holidays, and visits to gardens, and by other means.

and

> To found and maintain houses, for residence and for the carrying on of social and educational work in Ancoats and the surrounding districts, which shall become the common meeting ground of men, women and children of all classes, and where mutual goodwill, sympathy and friendship may be fostered.

and

> To ameliorate the conditions of life in Ancoats and other industrial districts of Manchester by direct effort and by influencing the

governing bodies of the city, and to create a higher sense of civic responsibility.[11]

The reports of the University Settlement, especially in its first decade, continue to indicate the range of work carried out, led by its joint wardens Alice Crompton from 1898, and T. R. Marr from 1902, to meet these criteria. What is most striking is the popularity and intensity of debate and discussion, and how the Settlement enabled and structured the relations between women of different classes who began to discuss women's rights. Among the activities of the various debating societies (the Fawcett, for women, and the Toynbee, for men and women), large numbers are mentioned, with over one hundred members and an average of 28 attendees at the Fawcett one year, which focused on issues pertinent to its members, hearing from Eva Gore-Booth on the subject of 'Men' at a 7 November meeting, and from her regular collaborators, Mrs S. Dickenson on 'Women's Trade Unions' on 30 January and Mrs E. A. Haworth on 'Economic Independence of Women' on 6 February. The 1901 report also mentions Eva Gore-Booth's leadership of the Elizabethan Society, aided by Christabel Pankhurst,[12] while the 1904 report on that society records that it was

> formed by Miss Eva Gore-Booth four years ago with the object of studying and trying to interpret some of the best dramatic literature [...] Lately, the society which meets once a week on Tuesday night, has somewhat widened its scope, and many interesting discussions on vital political and social questions have been held. The membership is necessarily small, but the attendance has been most regular.[13]

Eva Gore-Booth and Esther Roper

The Settlement was, as the reports show, a crucible for the forging of new ideas and new relationships. Eva Gore-Booth, an Associate member of the University Settlement, offers an example of the ways in which its openness refracted contemporary debates in impactful ways. Her path to Manchester and the Settlement is itself indicative of the changing times. Gore-Booth's background is well known, chiefly because of her appearance in William Butler Yeats's great poem about her family home, 'In Memory of Eva Gore-Booth and Con Markievicz', which takes as its subject the young poet's starstruck visit to the sisters' Big House at Lissadell, in Sligo; the poem also effectively conjoined her forever to her sister Constance, and the fame or notoriety that attended her as a revolutionary, an MP and a critic of the new Irish Free State. As her biographers Gifford Lewis and Sonja Ternan make clear, Gore-Booth's own achievements lapsed into the shadows, although she was herself a notably original and influential actor in other dramas of the age.

The Gore-Booth family owned a large estate in Sligo, where they were well regarded at a time when Anglo-Irish landlords were the subject of fierce post-Famine pressure. Sir Josslyn Gore-Booth set up the Drumcliffe Dairy Society in 1895, much influenced by the leaders of the Irish Co-operative movement, Horace Plunkett and George Russell (also known as AE, the name that he used for his work as a poet and painter). The family had ancient ties to Manchester, and an estate in Salford, where Gore-Booth's great-uncle Henry was rector of Sacred Trinity Church and also a member of the committee of the Industrial Schools. Her move to Manchester, however, was not connected to these family relations. Suffering from regular bouts of poor health, she was sent by her family to Bordighera, then a popular resort in Italy, to recover. It was during her convalescence there that she met the recently graduated Esther Roper, herself recovering from an illness, at Casa Corraggio, the house of the Scottish fantasy writer George MacDonald, which he had set up as an open house for culture and debate.

Esther Roper was one of Owens College's first women graduates, and completed her degree in English Literature and Latin and Political Economy at a time when Owens College was still adapting to having women as students, gradually allowing them to use the same building as their male peers. The women's undergraduate magazine *Isis*, some issues of which she co-edited, records her participation in debates, on equal rights in one case, as well as her successes in examinations and her graduation.

Roper's father grew up in Ancoats and left school at 11 to work in a factory, while her mother, Amanda Craig, was the daughter of

Irish immigrants, *her* father being a hackney inspector: Roper, speculates her biographer Gifford Lewis, may have been the first woman from a working-class background to receive a degree.[14] Her father had attended Sunday School in Ancoats and signed up to the Church Mission Society (CMS), which would eventually send him and, after his marriage, him and his wife to what is now Nigeria. There they learnt Yoruba and, in spite of difficulties including lengthy imprisonment, developed enough knowledge of Yoruba culture to present regular talks on the subject when Roper returned to Manchester. His relative eminence as a missionary seems to have allowed him to move upwards in Manchester society, or away at least from his Ancoats family. Esther boarded at a mission school in London while her parents were away, so she would have hardly known her father, who died in 1877. The CMS paid for her and her brother Reginald's education, as well as paying her mother's pension until she died, halfway through Esther's degree at Owens College, meaning, as Lewis notes, that she had to care for the family while completing her degree. As Sonja Tiernan points out, Roper's connection with the University did not end with the award of her degree. 'She kept ties with Owens College and became a leading member of their Social Debating Society. The Society had a female-only membership and organised many debates around issues of suffrage for women. In 1892 Roper became a graduate Associate of Owens College', an onerous process which required approval by the University's Senate and Council, and meant she was subsequently 'granted special privileges including the right to attend lectures, access to college facilities and a vote at court'.[15] After Roper began work at the North of England Suffrage Society, a heavy workload caused the illness which led to her Owens College lecturers sending her to Casa Corraggio, under the care of their friend MacDonald, in 1895.

Roper's meeting with Gore-Booth in Italy changed both of their lives, although its impact on Gore-Booth was more dramatic. She returned to Sligo initially and, inspired by Roper, established the Sligo Irish Woman's Suffrage and Local Government Association. She moved to Manchester to live with Roper in 1897 – they would share a terraced house in Heald Place in Rusholme for many years.

7.2 Esther Roper, 1919. Courtesy of Manchester Libraries, Information and Archives, M74081.

During the term of her involvement with the University Settlement, and with the larger trade union and suffrage campaign, she published four books of poetry with London publisher Longmans Green: *Poems* (1897), *Unseen Kings* (1904), *The One and the Many* (1904) and *The Three Resurrections* (1905). Each of these collections, along with her fifth collection, *The Egyptian Pillar* (1907), published in Dublin by Maunsel, is shaped by the enabling structures of the University Settlement, and may be usefully connected with that context, which itself placed such great faith in poetry's capacity for relief or even redress.

'In broken stammers speaks the inmost soul': poetry, gender and the limits of public speech

While Sonja Tiernan's biography has, as outlined above, generated considerable interest in Gore-Booth's unusual life trajectory, whose bravery and independence speaks so clearly to us, her poetry continues to be, for the most part, forgotten by readers, as well as by literary critics and historians. While individual poems are glossed for their subject matter (what they say about her sister's revolutionary experience,[16] or how she uses mythological settings to write about lesbian relationships,[17] or their mention of trade union activity),[18] her most cited poem remains 'The Little Waves of Breffny', an anthology piece whose dreamy twilit nostalgia seems, by its date of publication in 1904, a belated addition to a particular strain of Irish poetry:

> The grand road from the mountain goes shining to the sea,
> And there is traffic on it and many a horse and cart,
> But the little roads of Cloonagh are dearer far to me
> And the little roads of Cloonagh go rambling through my heart.
>
> A great storm from the ocean goes shouting o'er the hill,
> And there is glory in it; and terror on the wind:
> But the haunted air of twilight is very strange and still,
> And the little winds of twilight are dearer to my mind.
>
> The great waves of the Atlantic sweep storming on their way,
> Shining green and silver with the hidden herring shoal;
> But the little waves of Breffny have drenched my heart in spray,
> And the little waves of Breffny go stumbling through my soul.[19]

Esther Roper observed that Gore-Booth's 'most characteristically Irish poems were written in Manchester',[20] while AE commented of this one that 'you have slipped into it at last – the Celtic manner, as I always knew you could and would. It ought to be natural to you to do so – a West of Ireland woman, and County Sligo at that.'[21]

This, and many of the early poems, present the natural world (identified with Sligo) as the opposite of the northern city she lived in, and that opposition seems to duplicate the settings of her famous Sligo neighbour Yeats's 'Lake Isle of Innisfree'. However, in other poems, more grounded in her experience of city life in Manchester, the dynamic is very different to Yeats's, and to the Celtic school that AE was keen to promote. In 'The City', although it begins with a simple opposition, 'On through the iron day each stone-bound square / The soul of the green grass entombed hides', the poem's dynamic is more apocalyptic than nostalgic. Gore-Booth ends by directly quoting from Yeats's poem's last lines and imitating the characteristic triple-stressed molossus at his line's end ('While I stand on the roadway, or on the pavements grey, / I hear it in the deep heart's core'),[22] but her speaker departs from reverie to declare:

> Then does the spirit of the Wise and Fair
>   Break from her sepulchre and walk the town,
> The iron bonds are loosened everywhere –
>   No pavement grey can crush the green grass down.[23]

The faith in the natural world echoes the Settlement's Ruskinian (and Wordsworthian) faith in Nature's replenishing power; it might also be suggested that Gore-Booth's engagement with the political campaigns associated with her work at the Settlement, in relation to labour and especially to suffrage, enabled her to develop a poetics that is subtler and more original than her critics allow. The imaginative resource and command of different temporalities (the future and present, as well as the past) which mark her best poems is also what differentiates them from the more binary anti-colonial imaginary which shaped so much other Irish lyric poetry of the Celtic Twilight period.

Gore-Booth's experience, as an Associate member and Society leader at the Settlement, has been underplayed in the biographical representations of her poetry. In her entirely unnoticed monologue 'The Weaver', she challenges the representation of the child factory workers who were the regular constituents of the Settlement's open-door and open-garden policy (albeit the garden was also where the Round House's previous owner and his family were buried). The city's factory *loom* here is transformed by memory and imagination; intricate, surprising rhythms haunt the present moment:

> I was the child that passed long hours away
> Chopping red beetroot in the hay-piled barn;
> Now must I spend the wind-blown April day
> Minding great looms and tying knots in yarn.
>
> Once long ago I tramped through rain and slush
> In brown waves breaking up the stubborn soil,
> I wove and wove the twilight's purple hush
> To fold about the furrowed heart of toil.
>
> Strange fires and frosts burnt out the seasons' dross,
> I watched slow Powers the woven cloth reveal,
> While God stood counting out His gain and loss,
> And Day and Night pushed on the heavy wheel.
>
> Held close against the breast of living Powers
> A little pulse, yet near the heart of strife,
> I followed the slow plough for hours and hours
> Minding through sun and shower the loom of life.
>
> The big winds, harsh and clear and strong and salt,
> Blew through my soul and all the world rang true,
> In all things born I knew no stain or fault,
> My heart was soft to every flower that grew.
>
> The cabbages in my small garden patch
> Were rooted in the earth's heart; wings unseen
> Throbbed in the silence under the dark thatch,
> And brave birds sang long ere the boughs were green.[24]

A later poem, 'The Anti-Suffragist', echoes not Yeats but one of his key influences, Alfred, Lord Tennyson, and offers a rebuke to those who remain confined to their studies and whose work is not touched by a 'gleam of sunlight, / or a windowful of wind':

> The princess in her world-old tower pined
> A prisoner, brazen-caged, without a gleam
> Of sunlight, or a windowful of wind;
> She lived but in a long lamp-lighted dream.
>
> They brought her forth at last when she was old;
> The sunlight on her blanched hair was shed
> Too late to turn its silver into gold.
> "Ah, shield me from this brazen glare!" she said.[25]

This figure of the silent, sheltered woman recurs across Gore-Booth's poems, and there is an uncanny reach to how her poems characterise silence, twinning the child's ignored experience with the grown woman's knowledge of silence's engulfing power. If the silence is to be resisted, as was clearly the case in her activist work, it is also the case that her poems recognise and acknowledge it as a historical fact about the continuities across time of women's experiences and ideas. In 'The Elm Boughs', one of the fine poems in 1905's *The Three Resurrections* which takes up this idea, Gore-Booth's sense of the marginal precarity of her subjects is clear:

> The Elm boughs shudder in the sooty wind,
> From their bright leaves the City children know
> That somewhere the black world is good and kind,
> And through green woods the sunlit breezes blow.
>
> All starved and stunted from the poisoned sod,
> They shiver upwards through the stained air;
> These are the battered pioneers of God,
> Waving his green flag in the city square.
>
> Thus in the grey-built city of the mind
> Wave the green boughs of a few hostage powers,
> Their secret whispered to the soiled wind
> Holds all our faith in Beauty's austere flowers.
>
> Somewhere the fair and secret troops of spring
> Shine in strange colours icy clear and cold,
> But I pass on through dark streets wandering,
> Or dream a dream beneath the elm boughs old.[26]

In other poems, silence and the fate of being locked away or otherwise forgotten also emerge: in 'Andromeda' ('chained to the rock of this sheer world [...] through the dark secret spheres in silence glide[s]';[27] in 'Narcissus', where, far from the kingly world of deeds and flame and sword, 'the dark firs / dream on the edge of silence';[28] in the Irish immigrant woman imagined isolated in 'A Dweller by the Ocean' ('I who eat porridge from a wooden bowl / Whilst one dim candle gutters in the gloom / Do wonder at the greatness of the soul');[29] and in 'The Dreamer', whose shop life is vividly represented:

> When daylight dawns I leave the meadows sweet
> And come back to the dark house built of clay,

Over the threshold pass with lagging feet,
Open the shutters and let in the day.

The grey lit day heavy with griefs and cares,
And many a dull desire and foolish whim,
Leans o'er my shoulder as I spread my wares
On dusty counters and at windows dim.[30]

7.3 Eva Gore-Booth. Courtesy of the National Library of Ireland, Sheehy Skeffington Photographic Collection. Image altered (with permission) by Manchester University Press to remove mark.

The repeated stress on the women's silence surely originates in Gore-Booth's sense that women were, politically, silenced, even to the extent that the development of the Fawcett Debating Society at the Settlement was seen as necessary because of the ways in which women members of the mixed Toynbee Debating Society felt silenced or unable to speak in that context.[31] In the poems, this is not just a contemporary problem but a silence for which Gore-Booth listens in earlier eras too, as in the similarly gendered absence and silence of 'The Harvest of Silence' where 'Corn reaped in silence by silent reapers lay shrined in the innermost Shrine / Whilst outside in the darkness the Priestess fell tranced, and the god paced to and fro';[32] and in 'Immortalities', whose epigraph imagines 'a voiceless statue is Apollo' and:

> My secret treasure-house beyond the grave
> Holds but the stars of heaven, the gods of Greece,
> And some faint echo of the voice of Maeve.
> And the One Voice that is the Eternal Peace.[33]

In *The Egyptian Pillar* (1907), possibly her finest single volume, Gore-Booth returns to the idea of silence in a number of poems explicitly linking it to public speech. The book was written after the decisive breakdown of her friendship with Christabel Pankhurst, a friendship initiated at the Owens College Debating Society, but developed through their work together at the University Settlement. Gore-Booth and Roper opposed Pankhurst's resort to physical force in the suffragette cause, and her individual interest, as they saw it, in the limelight. In 'A Lost Opportunity', Gore-Booth writes,

> Others there were who spake with fire and art;
> I stammered, breaking down beneath the weight
> Of that great stone that lies upon my heart
> When with one passion all my nerves vibrate.
>
> Little I said, who had so much to say –
> That is the memory that sears and stings,
> My soul was on fire, my thoughts were clear as day,
> Yet had my soul no wings.
>
> No matter, when that force beyond control
> Sweeps on one side the cobwebs of the brain,

> In broken stammers speaks the inmost soul...
> Nor shall her passion smite the air in vain.[34]

Gore-Booth's silence is, then, applied in two separate spheres as she teases out where her hesitation comes from. In 'The Thriftless Dreamer', she imagines her home country under pressure from modernisers who have in mind a 'prosperous goal':[35] she defiantly addresses them, 'On through your anxious voices' fret and jar / The soul pursues unmoved her silent course'.[36] Here her (and Ireland's) silence is truer than any hasty or self-serving political plans.

Her nationality, and her relation to the nation, also inform another interesting poem in which Gore-Booth considers the bad fit between her ambitions and the public realm where she has become an activist and campaigner. Here she probes whether it is her nationality that might have prevented her from being fully able to commit to political action in England. 'The Street Orator' is one of a number of poems that seem to take their bearings from her reading of William Blake. In it, she recounts her experience as an activist in Lancashire, and how memories of Sligo accompanied her there, not always aiding her as she made her speeches:

> At Clitheroe from the Market Square
> I saw rose-lit the mountain's gleam,
> I stood before the people there
> And spake as in a dream.
>
> At Oldham of the many mills
> The weavers are of gentle mind;
> At Haslingden one flouted me,
> At Burnley all the folks were kind,
>
> At Ashton town the rain came down,
> The east wind pierced us through and through,
> But over little Clitheroe
> The sky was bright and blue.
>
> At Clitheroe through the sunset hour
> My soul was very far away:
> I saw Ben Bulben's rose and fire
> Shining afar o'er Sligo Bay.
>
> At Clitheroe round the Market Square
> The hills go up, the hills go down,

Just as they used to go about
A mountain-guarded Irish town.

Oh, I have friends in Haslingden,
And many a friend in Hyde,
But tis at little Clitheroe
That I would fain abide.[37]

The main thrust of another poem about public action returns not just to the impossibility of comprehensive public speech, but also to the motive and purpose of a political speaker, reflecting no doubt on Pankhurst and on the suffragettes' split from the suffragists. 'On the Embankment' examines how the pursuit of sectional political power offers only short-term gains and, as it brings historical silences within earshot, the poem, not altogether persuasively, presents this as a narrow escape for its unworldly speaker:

The Rich, the Great, the Wise are here, the Living and the Dead,
Where the Great Towers of Westminster hold the high heavens at
  bay,
And the poor souls who have no hope take fame or power instead,
Whilst many an obscure winged one goes smiling on her way.[38]

'Smoke and tumult': utopia and activism

Although Gore-Booth was critical of some aspects of Ruskin's work (his 'ideal of women was of course sentimental and impossible'),[39] her political philosophy, with its refusal of a purely material or simply economic basis, is clearly informed by the vision of his powerfully influential Rusholme lecture, imagining a political economy that valued education for all, and that understood that civic responsibilities could be part of both teaching and research, and the networks those activities could create in a city or national society. In that lecture, Ruskin famously segued from a close reading of a passage in Milton's 'Lycidas', focusing on the poet's strange image of the 'blind mouth' as a negation of what a leader ought to be (one who *sees*, one who *feeds* their flock), before he made his audience imagine a scene reported in the *Morning Post* earlier that year, in which a 58-year-old labourer had died from want of food at his home in Spitalfields: for Ruskin, it was Milton, and the ability to truly *read* his poem, that enabled him to see what had gone wrong with British society. The lecture's recommendation was to suggest that the best plan, in response to such terrible events, was that 'national libraries [should] be founded in every considerable city'.[40] A future is imagined where utopian visions of another world will inspire responsible leadership and progressive change.

The sense that art and poetry could play a part in making people see and understand the world around her remained vital to Eva Gore-Booth. Shaped by her experiences of debate and activism at the University Settlement, she would go on, alongside Roper and other colleagues she met in Ancoats, to be a significant organiser and unioniser of women workers in Manchester and nationally. Her sister, Countess Markiewicz, was radicalised by her visits to Manchester, taking part with Gore-Booth and Roper in, among other activities, a successful parliamentary by-election campaign against Winston Churchill. Her activism was not limited to England: Gore-Booth and Roper would, after the 1916 Rising, appeal to senior British politicians to commute the death sentence Markiewicz received for her part in the action. Gore-Booth's poems on this subject, and on her unsuccessful campaign to exonerate Roger Casement (the diplomat and Irish nationalist sentenced to death on charges of treason), are another notable chapter in this pacifist, vegetarian, feminist poet's biography. As her own health deteriorated, she and Roper moved to London in 1921, where she would die five years later. In 1928 Roper returned to Manchester for the installation of a stained-glass window to honour Gore-Booth's memory.

The *Irish Times* reported on the installation of this window, at the University Settlement's Round House, commissioned by Roper in memory of Gore-Booth, and made by the artist Ethel Rhind, a distant relation of Gore-Booth's, and described thus:

Under the rounded arch of neutral colours hawthorn trees heavily laden with white blossoms reach up and between the spaces one catches sight of the blue sky. Beneath the purple hills of the background a cornfield, half of which is cut and half uncut, runs down to a little lake. From here, a winding road sweeps to the foreground. Angels

7.4 Stained glass window by Ethel Rhind, formerly at the Round House, Ancoats. Commissioned by Esther Roper as a memorial to Eva Gore-Booth. Courtesy of Eberly Family Special Collections Library, Penn State University.

carrying spring flowers and conversing with one another walk on it. Rabbits frisk about and flowers spring up near a little pool, beside which an angel sits reading a book. It is an earthly Paradise.[41]

Its images recall the poem 'Peace' from Gore-Booth's third collection, *The One and the Many*, published in 1905, just as she disengaged from her University Settlement work, taking her experience there into her career defending workers' rights, pacifism and suffrage for women. That poem begins, 'I am sad with the city's sadness, sick of toil, / Choked with smoke and tumult, weary of noisy mills', but ends with images that speak to the ideas of art, spirit and the natural world as harmonious and offering a utopian challenge to the actual world around her:

> The great white daisies toss at ease in the long grass,
> I will fling down my soul to rest in this green glade,
> Where amongst waving fronds the silent angels pass,
> And brown hares fawn about their knees noiseless and unafraid.[42]

Notes

1 Janet Wolff and Mike Savage (eds), *Culture in Manchester: Institutions and Urban Change since 1850* (Manchester: Manchester University Press, 2013), pp. 3–4.
2 See Mark Crinson, *Shock City: Image and Architecture in Industrial Manchester* (London: Paul Mellon Centre for Studies in British Art, 2022), p. 138, for a detailed account of Ruskin's thoughts on 'Manchester devil's darkness'.
3 *Manchester Guardian*, 24 March 1904, p. 12.
4 *Manchester Guardian*, 24 March 1904, p. 12.
5 *Manchester Guardian*, 7 March 1900, p. 10.
6 Ibid.
7 P. J. Hartog, *The Owens College, Manchester (Founded in 1851): A Brief History of the College and Description of its various Departments* (Manchester: J. E. Cornish, 1900), p. 3.
8 Michael E. Rose, 'Voices of the people', in Alan Kidd and Terry Wyke (eds), *Manchester: Making the Modern City* (Liverpool: Liverpool University Press, 2016), p. 191.
9 Michael Harrison, 'Social Reform in Late Victorian and Edwardian Manchester with Special Reference to T. C. Horsfall', PhD thesis, University of Manchester, 1987, p. 156.
10 Mary D. Stocks, *Fifty Years in Every Street: The Story of the Manchester University Settlement* (Manchester: Manchester University Press, 1956), p. 33.

11 UML MUS/2/1, Annual Report of the University Settlement, 1902, p. 7.
12 MUS/2/1, Annual Report of the University Settlement, 1901, p. 23.
13 MUS/2/1, Annual Report of the Manchester Art Museum & University Settlement, 1904, p. 38.
14 Gifford Lewis, *Eva Gore Booth and Esther Roper: A Biography* (London: Allen and Unwin, 1988), p. 29.
15 Sonja Tiernan, *Eva Gore-Booth: An Image of Such Politics* (Manchester: Manchester University Press, 2012), pp. 35–6.
16 Laura Arrington, 'Liberté, Egalité, Sororité: the poetics of suffrage in the work of Eva Gore-Booth and Constance Markievicz', in Anna Pilz and Whitney Standlee (eds), *Irish Women's Writing 1878–1922: Advancing the Cause of Liberty* (Manchester: Manchester University Press, 2016), pp. 209–25.
17 Justin Quinn, *The Cambridge Introduction to Modern Irish Poetry 1800–2000* (Cambridge: Cambridge University Press, 2010), pp. 51–2.
18 Tiernan, *Eva Gore-Booth*, passim.
19 Eva Gore-Booth, *Selected Poems, With a biographical Note by Esther Roper* (London: Longmans Green, 1933), p. 44.
20 Ibid., p. 15.
21 Ibid., p. 17.
22 *Collected Works of W.B. Yeats*, ed. Richard Finneran, vol. 1 (New York: Scribner, 1997), p. 35. On Yeats's nationally inflected utopian imagination, see also 'Wherever men have tried to imagine a perfect life, they have imagined a place where men plough and sow and reap, not a place where there are great wheels turning and great chimneys vomiting smoke', quoted in Elizabeth Cullingford, *Yeats, Ireland, Fascism* (Basingstoke: Palgrave, 1984), p. 11.
23 Sonja Tiernan (ed.), *Eva Gore-Booth: Collected Poems* (Dublin: Arlen House, 2018), p. 155.
24 Ibid., pp. 164–5.
25 Ibid., p. 282.
26 Ibid., p. 205.
27 Ibid., p. 215.
28 Ibid., p. 195.
29 Ibid., p. 199.
30 Ibid., pp. 207–8.
31 '[I]t was noticeable that women, though always present as listeners at the Saturday [Toynbee Society] discussions, yet rarely if ever appeared as speakers [...] which led to the formation, in 1900, of a debating Society for women, bearing the esteemed name of Mrs Fawcett.' MUS/2/1, Annual Report of the University Settlement, 1900, pp. 30–1.
32 Tiernan (ed.), *Eva Gore-Booth: Collected Poems*, p. 196.
33 Ibid., p. 221.
34 Ibid., p. 232.
35 Ibid., p. 232.
36 Ibid., p. 233.
37 Ibid., pp. 228–9.

38 Ibid., p. 233.

39 'Women and the suffrage: a reply to Lady Lovat and Mrs Humphry Ward' (1908), in Sonja Tiernan (ed.), *The Political Writings of Eva Gore-Booth* (Manchester: Manchester University Press, 2015), p. 77.

40 John Ruskin, 'Of king's treasuries', in *John Ruskin: Unto This Last and Other Writings*, ed. Clive Wilmer (London: Penguin, 1997), p. 286.

41 *Irish Times*, 12 June 1928, p. 4.

42 Gore-Booth, *Selected Poems*, p. 59.

PART II

Civic university and civic decline

Stuart Jones

A familiar narrative depicts the twentieth century as a period of industrial decline and the erosion of civic institutions in Manchester and other northern cities. But recent historians have made the case for a different chronology, placing decline in the post-1945 period. The late Victorian idea of the city as the authentic locus of a citizenship rooted in place retained much of its vitality in the interwar period.[1] Economically, the cotton industry was certainly in irretrievable decline, but other sectors – notably engineering – were growing, and unemployment was consistently below the national average, unlike in the Lancashire mill towns to the city's north.[2] It was a period of major civic building and development projects, ranging from the Wythenshawe estate (the largest municipal housing estate in Europe) to the new Central Library.[3] The city's middle-class culture had a certain solidity that imparted its distinctive ethos to the University. In 1926, when Walter Moberly became vice-chancellor, it was (so his obituarist recalled) 'renowned for its enterprise in research and for the distinction of its scholars'. But it was under Moberly (1926–34) and his successor John Stopford (1934–56) that the University set about building a stronger relationship with the city, and especially with its business community, the aim being that the people of Manchester should 'look upon it as their own university'.[4]

The first half of the twentieth century was perhaps the University's golden age, if measured by its ability to attract academics at the very top of their field: Rutherford, Bragg and Blackett, successive holders of the Langworthy Professorship of Physics; Michael Polanyi in Chemistry and then (an *ad hominem* appointment) 'Social Philosophy'; Samuel Alexander in Philosophy; Tout, Powicke and Namier in History. The University's academics had cultural weight in the city's close-knit cultural and intellectual elite, not least because of the symbiotic relationship between the University and cultural institutions such as, notably, the *Manchester Guardian*.[5] C. P. Scott and his successors as editor, W. P. Crozier and A. P. Wadsworth, all drew copiously on Manchester academics as commentators and reviewers: the historians Lewis Namier and A. J. P. Taylor, the English professors C. H. Herford and H. B. Charlton, and the philosophers J. L. Stocks and Dorothy Emmet were all among those names (or, usually, sets of initials) that recurred time and again in its pages.[6] Indeed, as David Hayton points out in his study of three academic Zionists (Chapter 9), the *Guardian* was the one aspect of Manchester life beyond the University with which Namier did engage during his twenty-two year tenure of the chair of Modern History. It helped that the *Guardian* had a long-standing commitment to the Zionist cause, beginning in Scott's time but strengthened under Crozier's editorship (1932–44).[7]

In the 1930s Alexander and Namier were prominent supporters of the Academic Assistance Committee, which sought opportunities for academics fleeing Nazi persecution, and they had the backing of two successive vice-chancellors, Walter Moberly and John Stopford. It enabled the University to recruit international academic talent, though not without some resistance at a time of economic depression. The University's appointment of Michael Polanyi to the chair of Physical Chemistry in 1933 was a case in point. It demonstrated agility (something for which the University was not always known in the middle of the twentieth century), and it helped that Moberly as vice-chancellor had the strong backing of the University's treasurer, Sir Ernest Simon, whose ancestors were German Jews.[8] It brought to Manchester a man who was to prove one of the great polymaths of the twentieth century. But it also provoked a backlash in the press, where the University was lambasted for preferring a 'foreigner' over well-qualified native candidates.[9] No one else of Polanyi's stature was recruited by this means, but a steady stream of academic refugees enriched the University and the city in many ways. One such story is told here in Jonathan Aylen's brief account of the UMIST engineer Franz Koenigsberger.

In a remarkable book published in 1951 on the centenary of the foundation of Owens College, the Professor of English Literature, H. B. Charlton, powerfully evoked the 'Manchester idea' that had shaped the university where he had worked since 1912.[10] But this marked the end of the great age of the civic university, in Manchester and beyond. The bleakness of the city, rather than its dynamism, struck foreign visitors such as Michel Butor, who arrived as a tutor in French in 1951, and W. G. Sebald, who came to teach German fifteen years later, and Catherine Annabel has written a fascinating vignette showing how their fiction drew on their encounter with what was still a 'shock city', but perhaps now only in a negative sense. Meanwhile the University was much criticised by the younger generation of academics for professorial autocracy in academic matters and the inflated importance of lay officers in financial and strategic management.[11]

It was in the 1960s and 1970s that the English civic universities lost something of the sense of their distinctive mission. Both Manchester the city and Manchester the university seemed to be in decline. 'Less distinguished in arts than it was' was the laconic assessment of the University by an influential commentator in 1962.[12] That remark testified to the strength of the reputation that Manchester humanities had enjoyed in the days of Tout, Alexander and Namier, but it also conveyed a note of decline at a moment when the exciting things seemed to be happening at the new universities, which were (significantly) typically created and built in the form of out-of-town campuses. Some of them were named after counties or regions, a distinct departure from English (and indeed British) practice: Sussex, Essex, Kent, East Anglia. Manchester and other large civic universities haemorrhaged talented staff to these new universities, which now seemed to embody the future.[13] David Lodge caught the spirit of the time perfectly in his 1975 campus novel *Changing Places*, which told the story of a staff exchange between Rummidge University in the English Midlands and the State University of Euphoria ('Euphoric State') on the Pacific coast of the USA. Rummidge – loosely based on Birmingham, where Lodge taught –

> had lately suffered the mortifying fate of most English universities of its type (civic redbrick): having competed strenuously for fifty years with two universities chiefly valued for being old, it was, at the moment of drawing level, rudely overtaken in popularity and prestige by a batch of universities chiefly valued for being new.[14]

UMIST, however, does not quite fit a narrative of a declining university in a declining city. The Municipal College of Technology had had an unusual status in the first half of the twentieth century, as the only council-run institution that taught and examined at university level. In the 1950s it caught the national mood at a time when it was widely recognised that Britain under-invested in technology, and it acquired university rank (that is, central government funding via the University Grants Committee) as the Manchester College of Science and Technology. Ten years later, in a further boost to its standing, it was renamed UMIST, the University of Manchester Institute of Science and Technology. In the meantime, under a charismatic and ebullient principal, Vivian Bowden, it embarked on an ambitious programme of campus development, and one that explicitly aimed to be visible and open to the city. It was a bold restatement of the vision of the civic university, from an institution with undeniably strong local roots. Bowden was someone who thought that UMIST could only benefit from connecting itself creatively with Manchester's scientific and technological history; but while drawing copiously on that heritage, he also successfully constructed an enduring image of UMIST as a place of the future: a 'Crucible of Tomorrow's World' the *Guardian* called it in 1984.[15] But what gave UMIST a sense of drive and optimism was not so much its civic environment as the fact that it had been identified, alongside Imperial College and Strathclyde, as a UK *national* priority.[16]

For all UMIST's confidence that it embodied the 'white heat of technology', the sense that the University as a whole had lost something of its distinctiveness mirrored the impression that the city was also becoming just another northern city in an era of civic and industrial decline.[17] A. J. P. Taylor, who had roots in the region and taught at the University from 1930 to 1938, wrote a famous essay on Manchester in 1957. 'The merchant princes have departed', wrote Taylor (from his Oxford deer park). 'They are playing at country life in Cheshire or trying to forget Manchester in Bournemouth or Torquay.'[18] His essential point was right. The centenary of the award of city status was barely noticed in 1953, and historians have argued that the postwar welfare state, which largely supplanted voluntary and municipal agencies – rather than building on them, as Beveridge for the most part had envisaged – effectively sapped the roots of local citizenship.[19] When in Manchester Taylor had built up a close relationship with the *Manchester Guardian*, and he had remained close to Wadsworth, editor from 1944 to 1956. But soon after Wadsworth died, and Taylor wrote that article, the *Guardian* dropped 'Manchester' from its title (1959); in 1964 its editorial staff moved to London; and in 1972 it vacated its Manchester offices altogether. The journalist Andy Spinoza, who arrived in Manchester as an undergraduate in 1979, recalls the impression that the city was 'sliding into the dustbin of history'.[20]

Notes

1 Charlotte Wildman, 'Urban transformation in Liverpool and Manchester, 1918–1939', *Historical Journal*, 55:1 (2012), 119–43; Charlotte Wildman, *Urban Redevelopment and Modernity in Liverpool and Manchester, 1919–1939* (London: Bloomsbury, 2016), ch. 1; Tom Hulme, *After the Shock City: Urban Culture and the Making of Modern Citizenship* (Woodbridge: Boydell, 2019).
2 Wildman, 'Urban transformation', 123.
3 Ibid., 128–9, 134–5.
4 'Sir Walter Moberly. Educational statesman', *The Times*, 2 February 1974, p. 14.
5 On the relationship between the University and the *Manchester Guardian*, see Mary Stocks, *My Commonplace Book* (London: Peter Davies, 1970), p. 152. I owe this reference to John Ayshford.

6 So was Mary Stocks (John Stocks's wife), who was a significant figure in both civic and academic communities.

7 Daphna Baram, *Disenchantment: The Guardian and Israel* (London: Guardian Books, 2004), ch. 2.

8 On this family, see John Ayshford et al. (eds), *The Simons of Manchester: How One Family Shaped a City and a Nation* (Manchester: Manchester University Press, 2024).

9 Bill Williams, *Jews and Other Foreigners: Manchester and the Rescue of the Victims of European Fascism* (Manchester: Manchester University Press, 2013), p. 39.

10 H. B. Charlton, *Portrait of a University, 1851–1951* (Manchester: Manchester University Press, 1951).

11 Brian Pullan with Michele Abendstern, *A History of the University of Manchester 1951–73* (Manchester: Manchester University Press, 2000), ch. 8.

12 Anthony Sampson, *Anatomy of Britain* (London: Hodder and Stoughton, 1962), p. 209.

13 Pullan, *History of the University of Manchester*, pp. 98–9.

14 David Lodge, *Changing Places* (1975) (Harmondsworth: Penguin, 1978), p. 14.

15 'Crucible for tomorrow's world', *Guardian*, 27 November 1984, p. 18. But for a different note, see the *Guardian*'s review of *Artisan to Graduate* in 1975: 'The relationship with local industry which from its humblest beginnings gave the institution its strength is now providing in one way its weakness as particular industries in the North-west go into decline. So although in one sense this is a celebratory book, it is not one with a happy ending.' 'Made in Manchester', *Guardian*, 18 January 1975, p. 19.

16 On UMIST campus development I am indebted to forthcoming work by Richard Brook.

17 'The white heat of technology' was the famous phrase used by Harold Wilson in his party conference speech of 1963. It is often taken to sum up the spirit of his 1964 general election campaign. In October 1964 he appointed the recently ennobled Lord Bowden as Minister of Education and Science. On Bowden I am indebted to forthcoming work by Erin Beeston.

18 A. J. P. Taylor, 'Manchester', *Encounter*, 8:3 (March 1957), 13.

19 Hulme, *After the Shock City*, pp. 203–8; Jose Harris, *William Beveridge: A Biography*, 2nd edn (Oxford: Clarendon Press, 1997), pp. 451–61.

20 Andy Spinoza, *Manchester Unspun: Pop, Property and Power in the Original Modern City* (Manchester: Manchester University Press, 2023), p. 1.

Catherine Chisholm and the Manchester Babies' Hospital

Peter D. Mohr with Stuart Jones

Catherine Chisholm (1878–1952) was the first woman to graduate in medicine from the University of Manchester. She went on to play an important role in the creation of modern neonatology. The Manchester Babies' Hospital (MBH) – founded in 1914, and renamed in 1935 the Duchess of York Hospital for Babies – was her creation, and it deserves to be remembered as a remarkable piece of social enterprise. Building on the example of the London Infants' Hospital founded by Theophilis Kelynack in 1903, it filled a glaring medical need: nationally, there was growing concern at the persistently high infant mortality rate when the overall mortality rate was falling. In 1914 around 13 per cent of Manchester babies died in their first year.[1] The MBH was also almost entirely the creation of women, run by women, and staffed by women, at a time when women doctors were effectively excluded from hospital positions. The story of its creation tells us something important about the agency of networks of educated women – professionals, volunteers and elected representatives – on the eve of the First World War.

The daughter of a GP in Radcliffe (near Bury), Chisholm studied Classics at Owens College, graduating in 1898, and then entered the Medical School as its first woman student, graduating with a First in 1904. She set up in general practice on Oxford Road, providing a much-needed service for local women, children and female university students; but she soon realised that her real vocation was the medical

V2.1 Dr Catherine Chisholm on the occasion of the award of the CBE, 1935. Courtesy of University of Manchester, Museum of Medicine & Health.

care of children and the growing infant welfare movement. She was medical advisor for the Santa Fina Society, part of the University Settlement in Ancoats, which supported disabled ('crippled') children, and in 1908 she was appointed as medical officer for the Manchester High School for Girls, then located on Dover Street, in the immediate vicinity of the University. In the same year she became honorary physician to the Chorlton-on-Medlock Dispensary for Women and

Children: the first woman to be appointed to a senior medical post in Manchester. Her efforts to obtain an established hospital position were frustrated, however: the Manchester Royal Infirmary refused to employ women doctors, and she was turned down for the post of honorary consultant children's physician at the St Mary's Hospital for Women, which opened a new branch on Oxford Road in 1911.

It was around 1910 that Manchester's Medical Officer of Health, James Niven, introduced Chisholm to Margaret Ashton, who in 1908 had been elected to Manchester City Council as its first woman councillor. Ashton was the first chair of the council's Infant Life Protection Sub-Committee – later the Maternity and Child Welfare Committee (MCWC) – and Chisholm was co-opted on to the sub-committee to advise on maternity, infant welfare clinics, schools for mothers, municipal nurseries and health visitors. It was the collaboration between Chisholm and Ashton that made the Babies' Hospital possible.

Ashton was the daughter of Thomas Ashton, the cotton manufacturer and merchant who appears earlier in this book as the 'second founder' of Owens College in the early 1870s (Chapter 3). He left an estate worth over half a million pounds gross when he died in 1898, which made him one of the wealthiest of Manchester's cotton masters. Margaret was one of nine children, but her inheritance was sufficient to make her independently wealthy. She also inherited something of her father's sense of public duty and the responsibilities of wealth. It was she who provided the financial backing that launched the Babies' Hospital: half of its running costs in its first year, to start with, which gave it a solid platform from which to seek a wider range of subscribers. She was not in a position to bankroll the hospital, but alongside her money she contributed civic influence and strength of purpose.[2]

The MBH was the second hospital in the UK specifically for babies. Unlike the London Infants' Hospital, it was staffed almost exclusively by women doctors. It was also managed largely by women: six women doctors served on the management committee, alongside three representatives of the city council (including Margaret Ashton) and representatives of the subscribers. This last group were mostly women from business, professional and academic families: they included, notably, Shena Simon, Janet Zimmern (both Liberal city councillors in the 1920s) and Mary Tout. Shena Simon was the first chair of the management committee. She and her husband Ernest were notable benefactors of the hospital, especially via the two Simon family engineering businesses. All three of these women had University connections as members of the Court of Governors (as was Chisholm), and, in Simon's case, as a long-standing member of the University's Council. The Touts were closely involved with the University Settlement, as was Zimmern, who (as Janet Blair) first came to Manchester to work at the Settlement, and was appointed head of its Women's House in 1911. She may well have encountered Chisholm first through the Santa Fina Society. Mary Tout had studied at Owens College at the same time as Chisholm and married the historian Thomas Tout. Like Chisholm, she was a founder member of the British Federation of University Women (now the British Federation of Women Graduates), which was founded at a meeting at the Manchester High School for Girls in 1907.[3] The BFUW was to be active in campaigning to widen professional opportunities for women. It was on behalf of the BFUW that Chisholm and Tout co-authored a letter to the *Manchester Guardian* in 1909 protesting against the Royal Infirmary's continuing insistence that it was unable to appoint women to medical positions.[4] The establishment of the Babies' Hospital as a women-run institution has to be understood in this context.[5]

Another network that overlapped with the BFUW was the Manchester and Salford Women's Citizens' Association (WCA), a cross-party group formed in 1914 to help women gain more influence in local government, especially on social problems affecting women's lives in particular. Chisholm, Ashton, Tout, Simon and Zimmern were all active members. Although the WCA had limited influence on local government policy, it was useful to Chisholm: a WCA sub-committee under her would report to Margaret Ashton, chair of the MCWC, on the needs of the MBH, and Ashton would in turn advise the city council on the allocation of funds to support the MBH, as part of public health policy. This was important: the MBH

V2.2 Ernest and Shena Simon (Lord Mayor and Mayoress) visiting the Manchester Babies' Hospital in 1922. They were notable benefactors of the hospital, and Shena Simon was the first chair of its management committee. Courtesy of University of Manchester, Museum of Medicine & Health.

was a 'voluntary' hospital, principally reliant on its subscribers, but from 1916 most of its beds were funded by the council.

The MBH had no formal connection with the University, and its history is much more a civic than a University history; however, the University was an important part of the context in which Chisholm and the hospital worked. Medical students were taught at the MBH, and Aron Holzel, lecturer in Child Health, did important research into metabolic diseases there. John Stopford – Professor of Anatomy from 1919, dean of the Medical School from 1923, and vice-chancellor from 1934 – was a close friend of Chisholm's and an influential ally, as was his wife, Lily Allan, who served as ophthalmic consultant at the MBH. Chisholm was the University's medical offer for women students, and lectured on vaccination.

The MBH started in premises in Chorlton-on-Medlock and then in Levenshulme, both close to the University, before settling in Cringle Hall, Burnage. By the 1920s the MBH was admitting over 600 patients a year, and the construction of a surgical wing in the 1930s realised Chisholm's vision of a complete children's hospital. The Duchess of York – later Queen Elizabeth the Queen Consort – visited in 1935 to open the surgical wing, and rename the hospital the Duchess of York Hospital for Babies. In the same year Chisholm was appointed CBE for her work for child health. The MBH had a distinctive place in the complex map of hospital provision before the establishment of the National Health Service, but its early history also illuminates networks of influence by which women could shape social institutions in early twentieth-century Manchester.

Notes

1 Peter D. Mohr, 'Women-run Hospitals in Britain: A Historical Survey focusing on Dr Catherine Chisholm and the Manchester Babies' Hospital', PhD thesis, University of Manchester, 1995, p. 207; Peter D. Mohr, 'Dr Catherine Chisholm (1879–1952) and the Manchester Babies Hospital', *Transactions of the Lancashire and Cheshire Antiquarian Society*, 94 (1998), 95–110.
2 Lady Simon of Wythenshawe [Shena Simon], *Margaret Ashton and her Times* (Manchester: Manchester University Press, 1949).
3 'University women's federation returns to birthplace: jubilee meeting in Manchester', *Manchester Guardian*, 13 July 1957, p. 3.
4 'Women doctors and the Royal Infirmary', *Manchester Guardian*, 2 February 1909, p. 14.
5 On these networks, see Mohr, 'Women-run Hospitals', ch. 5.

8

Three Zionists: Samuel Alexander, Chaim Weizmann and Lewis Namier

David Hayton

Speaking at a public dinner in 1951, Eliahu Eilath, the Israeli ambassador to the United Kingdom, referred to Manchester as 'one of the most important cities in the growth of the Zionist movement' at the beginning of the twentieth century.[1] According to Eilath, this importance derived principally from the presence of Chaim Weizmann (1874–1952), sometime Reader in Chemistry in the University, the first president of the state of Israel and the man who had persuaded the British Foreign Secretary Arthur Balfour to make his famous declaration in 1917 of the British government's support for the establishment of a national homeland for the Jewish people in Palestine. When Weizmann had arrived in Manchester in 1904 as a research fellow, Zionism was already a force in the city, but his personality attracted and energised a rising generation of activists and intellectuals, men such as Leon Simon and the brothers-in-law Simon Marks and Israel Sieff, all of whom had been educated at Manchester Grammar School (and, in the case of Sieff, at the University). As Eilath observed in his speech, the support given by the *Manchester Guardian* had also played a significant role in mobilising liberal opinion in favour of the Zionist cause, and here again Weizmann's influence can be detected in inspiring the *Guardian* journalist Harry Sacher, who had also married into the Marks family and, like his brothers-in-law, came under Weizmann's spell.

The University's significance in the history of Zionism, however, rests not just on Weizmann's shoulders. Lewis Namier (1888–1960), Professor of Modern History from 1931 until 1953, had served as political secretary of the Zionist Organisation during a particularly severe crisis in Palestine in 1930, and while he held no official post in the organisation during his time at Manchester he remained active in the Zionist movement until the end of the Second World War. Another Jewish intellectual from a slightly older generation, Samuel Alexander (1859–1938), Professor of Philosophy from 1893 to 1924 and a towering figure in Manchester's intellectual life, also exhibited sympathy for the Zionist cause, albeit in a less obtrusive way; not in political activism or grand gestures but through offering practical assistance to the Jewish community in Palestine, principally the Hebrew University that had been established in Jerusalem, and above all by a willingness to assist fellow Jews in distress, especially those fleeing persecution in Europe during the 1930s.

These three men differed sharply in background and personality, and each reacted to the city and the University in different ways. Alexander, who gave up an Oxford college fellowship for his chair, wholeheartedly embraced Manchester life. Weizmann at first disliked his new environment, but eventually found an entrée into middle-class society and encountered like-minded individuals. His relationship with the University, by contrast, began well, only to sour when he quarrelled with the Professor of Chemistry, William Perkin, and was passed over for the chair when Perkin moved to Oxford. Namier experienced a similar disillusionment after what was for him a surprising and very welcome appointment – in fact, his first full-time academic post. For Namier, Manchester as a city held few charms, and during his twenty-year occupation of the modern history chair he based himself in London and commuted by train for only a few days each week.

Samuel Alexander

Samuel Alexander not only belonged to an older generation than Weizmann or Namier, he had no direct experience of the travails of European Jewry which provided the stimulus and essential rationale behind the ideology of Zionism.[2] He was born into a family of liberal Jews in Sydney, the offspring of an Australian father who died when Alexander was young, and a South African mother, left with more than adequate means after her husband's death. There was enough money for young Samuel to be educated at the recently established Methodist boarding school, Wesley College in Melbourne, and at Melbourne University. He secured a scholarship at Balliol College, Oxford, then at the height of its reputation under the mastership of Benjamin Jowett, took a first in Greats, won the Green Prize in Moral Philosophy, and in 1893 became a fellow of Lincoln College, the first practising Jew to be elected a fellow of an Oxbridge college. He published his Green Prize essay in 1889, heavily influenced by the Idealist school then dominant at Oxford, but became attracted by developments in evolutionary biology and experimental psychology and began to take his own work, and the subject as a whole, in a very different direction.

Alexander suffered no setbacks or difficulties in his academic career. Despite disappointing some of his admirers by publishing relatively little, he was elected a Fellow of the British Academy, and in 1930 was appointed to the Order of Merit. His endearing eccentricities, added to his renown as a philosopher, made him a popular figure across the city of Manchester, and after retiring in 1924 he was happy to stay there for the rest of his life.

Throughout his time in the University Alexander espoused progressive political causes, notably women's suffrage (though he was characteristically sceptical about some of the tactics employed by suffragettes). He also played a full part in supporting the activities of Jewish groups in the University, especially the Jewish Students' Society. His attitude to the Zionist project was more complex. While there can be no doubt that he was broadly sympathetic, he chose not to exploit his national, or even his local, fame to advance the aims of the Zionist Organisation. Perhaps he lacked the emotional drive so evident in both Weizmann and Namier, having himself enjoyed a comfortable upbringing far from the traumas experienced by East European Jewry. A good friend to Weizmann during the

8.1 Samuel Alexander in 1917, photograph by Walter Stoneman. Courtesy of the National Portrait Gallery.

struggles over the professorship of Chemistry, he aroused Weizmann's ire in 1917 when, as co-president of the Conjoint Committee on Foreign Affairs, a body devoted to the defence of the rights of European Jews, he joined the anti-Zionist Claude Montefiore in issuing a statement rejecting Weizmann's proposal to establish a Jewish commonwealth in Palestine under British protection.[3] As quick to forgive as to anger, Weizmann did not allow this incident to mar their friendship and some five years later wrote to urge Alexander to make a public statement in favour of a Zionist solution to the Jewish question, at a time when, as Weizmann said, the movement was 'being attacked from many quarters'.[4] Even though no statement was forthcoming, the two men remained on good terms, and during the intense political negotiations in 1931 between the Zionist Organisation and the British government over the scale of Jewish migration to Palestine, Alexander sent Weizmann a private message of support.

Tributes flooded in after Alexander's death in 1938. Among others, Namier wrote an appreciation for the *Manchester Guardian* which described Alexander as 'a good Jew and a Zionist' who 'felt deeply with his people'.[5] Namier also persuaded the vice-chancellor of Manchester and the master of Balliol to join in an appeal for a commemorative fund, writing again that 'Samuel Alexander was a Jew who felt strongly with his people'.[6] Although Namier mentioned generous subscriptions 'for Zionist purposes', and Alexander's steadfast support for the Hebrew University, the practical outcome of these strong feelings was to be sought primarily in efforts to assist refugees from Nazi oppression.

Chaim Weizmann

In contrast to Alexander, Chaim Weizmann's early life exposed him directly to the worst perils threatening his race.[7] A typical *shtetl* Jew, he was brought up in a small settlement in 'White Russia' (modern-day Belarus) in a strongly observant family, which embedded in him an intensely emotional attachment to the people to whom he was to dedicate his life. He also spent part of his education in Germany, first in Darmstadt and then in Berlin, forcing him to witness at first hand the rising tide of antisemitism in European public life, and the superciliousness of assimilationist German Jews whose open disdain for *Ostjuden* fuelled his own more radical response to what was coming to be perceived as the 'Jewish question'. The influence upon him of the writer Asher Hirsch Ginsberg, who published under the pseudonym Ahad Ha'am ('one of the people'), was also evident in the emphasis Weizmann placed on the spiritual regeneration that Zionism would accomplish, which he regarded as of equal importance with the historical duty of achieving nationhood and the practical necessity of enabling Jews in Tsarist Russia, in particular, to escape the increasingly vicious pogroms to which they were being subjected.

The Zionist movement which Weizmann joined had emerged in the late nineteenth century as a response to increasing antisemitism across Europe: a renewed wave of persecution from the 1880s onwards within the 'Pale of Settlement' in Russia; and in France the explosion of the Dreyfus affair in 1894. Across the continent, the rise of nationalism focused hostile attention on Jewish communities as anomalous and even as alien elements within emerging nation-states and created the so-called 'Jewish question'. The same ideological impulse in turn prompted Jewish intellectuals to make their own response to the emphasis on nationhood by calling for the establishment of a nation-state in what they regarded as the historically Jewish lands of Palestine, 'Eretz Israel' (the land of Israel), then under the rule of the Ottomans. This objective, of course, ran counter to the aspirations of liberal Jews, particularly in Germany, who believed that emancipation could be achieved by assimilating into the societies and cultures in which they lived. The most important opponent of such assimilationist doctrine was the Hungarian journalist Theodor Herzl, whose pamphlet *Der Judenstaat* (The Jewish State), published in 1894, argued that Jews could only avoid religious intolerance and persecution by removing themselves to Palestine. Herzl's brand of Zionism was not the only version available, and was indeed subject to the challenges of younger critics, but the main body of activists, including Weizmann, began by looking to Herzl for intellectual leadership.

At successive Zionist congresses from the late 1890s onwards, Weizmann quickly established a reputation as a firebrand. It was not just assimilationist Jews who felt the lash of his tongue, or those socialists who looked for other kinds of political solution to the 'Jewish question'. Influenced by Ahad Ha'am, he also began to question Herzl's concentration on making diplomatic approaches to European heads of state. He was not opposed to this strategy in principle, and indeed would spend much of his later career as the leader of the Zionist Organisation in attempting to establish a Jewish state through the intervention of the British mandatory authority in Palestine. Rather, he now considered Herzl's efforts ineffectual, and thus as a distraction from what he himself regarded as the essence of the movement. For Weizmann this was embodied in 'the victorious force of the Idea' which he believed was destined to triumph.

Weizmann thus came to Manchester with deeply ingrained beliefs to which he was accustomed to give forceful expression. This did not make him an easy colleague in whatever context he found himself. On the positive side, he had also begun to make a reputation for himself as a scientist, and could boast of promising discoveries while working in Berlin, even if these had not yielded the commercial profits he had envisaged (a failure which, of course, he turned into a personal grievance). His potential was certainly recognised, but he had not achieved enough to warrant more than the lowliest research appointment, as an assistant to the Professor of Chemistry, William Perkin.

At first Manchester held no attractions for him, and he regarded the bulk of the city's Jewish population as 'a very stupid rabble', unsophisticated and utterly materialist.[8] But gradually things improved: financially, as he supplemented his University salary with an income from work with local industrialists; academically, as his work with and for Perkin moved in new and exciting directions; and socially, as he made the acquaintance of middle-class Jews, including politically sympathetic individuals. Weizmann also became involved in local Zionist organisations, and in 1905 was sent to the seventh Zionist Congress, in Basle, as Manchester's delegate, enabling him to re-enter Zionist politics at an international level.

The decade he spent in Manchester from 1904 to 1915 enabled Weizmann to re-energise his political career, and brought him financial security and even a modest prosperity. It also provided him with the resources and intellectual stimulation needed to develop his academic work. He published frequently, and began to explore different aspects of his subject. But his more than ample *amour*

8.2 Chaim Weizmann in his laboratory at the University of Manchester, c. 1913. Courtesy of the University of Manchester.

propre, his volatile temperament, and his clumsiness in intrigue, all of which would be very obvious during the turbulent years when he was at the head of the Zionist Organisation, inevitably resulted in conflict and isolation within the University. He quarrelled decisively with Perkin, and when the professor was elected to a chair at Oxford in 1912, Weizmann, who felt he had more than earned the right to succeed, found himself passed over in favour of

Perkin's brother-in-law. The award of a readership he did not see as appropriate compensation. The whole experience tainted his view of Manchester, and although he maintained his scientific interests and stayed on at the University until seconded to London in 1914 for war work (related to the development of explosives) he did not return when the First World War came to an end and instead devoted himself to Zionist affairs on the international stage.

Lewis Namier

Like Alexander and Weizmann, Lewis Namier had no prior connection with Manchester before his arrival in 1931.[9] For him, the landscape of a northern English industrial city was quite alien and required a sharp adjustment in his expectations. As a child he had been cosseted in a comparatively wealthy environment, and his experience of urban life before his arrival in England in 1908 had been of Habsburg imperial cities, large and small, distinguished by a *fin-de-siècle* elegance.

Namier's upbringing in Eastern Europe had been very different from Weizmann's, and in many ways uncongenial to the development of a Zionist outlook. He was born Ludwik Bernstein into a family which was strongly assimilationist. His father Jozef, a lawyer and would-be landed gentleman in the Habsburg province of Galicia, considered himself first and foremost a Pole. There was little religion in the Bernstein household, certainly no observance of Jewish rites and customs. Eventually, at the end of the First World War, the family's Polish self-identification culminated in Namier's parents and sister converting to Roman Catholicism and readopting the former family surname of Namierowski. For Namier himself, however, this was not an option to be considered. Friends observed his indifference to religion and indeed a pronounced anti-clericalism, which encompassed rabbis as well as priests. When he married for the second time in the 1940s, his wife, herself a Russian Orthodox mystic, insisted he be baptised, so he chose the Church of England as representing the most politically innocuous form of Christian faith available.

Ever the contrarian, even at an early age, Namier began by rejecting his father's earnest liberalism in favour of a brand of socialism. He was also drawn towards a pan-Slavist outlook, seeking inspiration in Russian history and literature, especially the novels of Dostoyevsky, which he regarded as truly representative of the soul of Slavdom. Unlike other East European Jews he had little time for German culture, and no love for the Habsburgs, whom contemporaries such as the novelist Joseph Roth regarded as the protectors of the empire's Jewish subjects. Namier's first experience of university education, at Lemberg (modern-day Lviv), complicated these early allegiances and also brought home to him the peculiarity of his own background. The socialist parties were divided and mutually hostile, and he had nothing in common with any of the Jewish groups in that city: he had grown to despise the timidity of assimilationist Jews, whom he would later characterise as 'The Order of Trembling Israelites', while all the various Zionist factions were culturally alien.[10]

Namier's time at Lemberg was cut short, and after various false starts he fetched up at Balliol College, Oxford, which he came to regard as his intellectual home (and where Samuel Alexander had been a student more than twenty years before). He adopted the intellectual prejudices of the university in general, and his college in particular, idealising the British empire as a beneficent political and social force quite unlike the European empires with which he was familiar, and adopted the progressive social democrat ethos of his teachers and many of his friends. Eventually he acquired British nationality and changed his name, adapting Namierowski into a form that he thought, mistakenly, was quintessentially British. At first glance there was little indication of the Zionist that he would become. The one attempt he made to penetrate a specifically Jewish coterie at Oxford ended when a visit to the Adler Society left him aware of how little he had in common with observant Jews, even the public-school-educated Jewish undergraduates at Balliol.

Yet there were elements in Namier's character and thinking that prepared him to accept Zionism. In particular, he believed strongly, as his contemporaries did, in the importance of race as a determinant

Plate 7 The bluedot festival at Jodrell Bank. Courtesy of One Eye.

Plate 8 The Post-Crash Economics Society at the University of Manchester. Courtesy of Jon Super/*The Guardian*/eyevine.

Plate 5 The Turing memorial, in Sackville Gardens, Manchester. Photographed by KGGucwa.

Plate 6 Brian Cox, who 'leapt to ubiquity' in 2008. Courtesy of the University of Manchester.

Plate 4 Portrait of Sir Henry Roscoe by Hubert von Herkomer. Roscoe was the pre-eminent academic figure at Owens College, where he was Professor of Chemistry, 1857–86. Courtesy of the University of Manchester.

Plate 3 The Oxford Road elevation of Owens College, completed around 1887. Courtesy of the University of Manchester.

Plate 2 Dr George Birkbeck, one of the founders of the London Mechanics' Institution (now Birkbeck, University of London). Courtesy of Birkbeck, University of London.

Plate 1 Thomas Turner, surgeon, founder of the Pine Street School of Medicine, 1824. Source: Wellcome Collection.

Plate 9 External view of the Nancy Rothwell Building which forms the Manchester engineering campus development. Courtesy of the University of Manchester.

Plate 10 Illustrative image of the proposed new square at ID Manchester, on the former UMIST campus. Courtesy of ID Manchester Joint Venture.

Plate 11 Visualisation of the the remodelling and regrassing of the Old Quadrangle, undertaken in 2024 to mark the University's bicentenary. Courtesy of the University of Manchester.

Plate 12 University of Manchester's Light Up event, celebrating the start of its 200th year. Photograph by Monica Naughton Crimmins.

of human behaviour and a necessary element of statehood. He also recognised, indeed emphasised, the part played by religion as a marker of national identity, and in influencing national character and ideas. Just as, in his view, the Roman Catholicism of Polish nationalists predisposed them to unthinking obedience to an absolutist political system, the devotion to individual liberty and representative government shown by the English (and the first American colonists) could be traced to the legacy of English Puritanism. But while admiring Calvinist Protestantism, Namier was also becoming more and more aware of his Jewishness. Unlike the middle-class English Jews who thrived in Edwardian Oxford, Namier stood out from his contemporaries because of his European origins and heavily accented spoken English. A polite form of antisemitism was endemic in 'Varsity' life. Namier did not attribute his failure to secure an All Souls' prize fellowship to prejudice, as Weizmann blamed slowness of promotion at Manchester on the fact that he was a Jew, but the sentiments of those around him cannot have entirely escaped his cognisance: even his greatest supporter, his tutor A. L. Smith, began a testimonial for him with the words, 'I started on him with the usual prejudice against Jews.'[11]

Namier's induction into the Zionist movement during the 1920s had come about largely through contact with Weizmann, with whom he felt a strong personal bond, possibly even a transfer of filial attachment following his estrangement from his own father, who died in 1922. To some extent Namier's relationship with Weizmann paralleled that previous experience, being a mixture of idolatry and exasperation. It would seem likely that the strongest force driving him towards Weizmann and Zionism was his experience during the First World War and its immediate aftermath. Working in the Political Intelligence Department of the Foreign Office, he was able to read confidential despatches and a range of European newspapers, which revealed the depth of antisemitism among nationalist political parties in Eastern Europe, and the detail of outrages committed against Jewish populations. The way that his own reports were ignored by Foreign Office mandarins also brought home the pervasiveness of antisemitism in the British establishment, and although there were

8.3 Lewis Namier in the 1940s, photograph by Elliott & Fry. Courtesy of the National Portrait Gallery.

many who stood by him, including his immediate superiors, he had to bear with innuendoes and even outright persecution in the right-wing press, and questions in Parliament about this 'Austrian' Jew in the Foreign Office who was undermining loyal Polish allies. Disillusionment was completed by what he perceived as Britain's failure to prevent both the persecution of Jews in the new Polish state after 1918, and the onward march of a form of 'Polish imperialism' on the country's eastern borders, at the expense of the Ukrainians, who had another claim on his loyalty. He left the Foreign Office in 1920 while friends and colleagues stayed on, and turned his attention to writing history, and to Zionist politics.

Namier had first met Weizmann and other leading Zionists around 1919, and gradually drew closer and closer to the man he came to call 'the chief', attending Zionist congresses in the 1920s, and beginning to publish articles explaining and defending the precepts of Zionism. The most important of these, published in the *New Statesman* in 1927, which made a particular impression on the young Isaiah Berlin, argued the case for establishing a 'national home' for Jews in Palestine partly on practical grounds – the unhappy consequences for Orthodox Jews in Russia and Eastern Europe forced to live under oppressive and intolerant regimes – and the overriding principle that 'every nation must somewhere have its own territorial centre'. The Balfour Declaration and the Palestine Mandate offered an unprecedented opportunity, since the British were 'the best [friends] we have had since Cyrus and Alexander [the Great]'.[12]

Namier's experience of the Zionist Organisation was not, however, a happy one. At Weizmann's insistence, he agreed to take on the office of political secretary in 1929 at a time of great crisis in Zionist affairs. The Zionist Organisation had been encouraging the settlement of Jewish immigrants in the mandated territory, but increasing tensions between Jews and Arabs, particularly in Jerusalem, had erupted into communal violence. The Labour government in Britain established a commission of inquiry, which blamed excessive Jewish settlement. Eventually the Colonial Secretary Sidney Webb, Lord Passfield, published a White Paper recommending tight controls over Jewish immigration and Jewish acquisition of land. Namier found himself engaged alongside Weizmann and David Ben-Gurion in prolonged and tense negotiations with government ministers, which resulted in the withdrawal of Webb's proposals.

Rather than celebrating this success, enemies within the Zionist Organisation questioned Weizmann's strategy, accusing him of having 'sat too long at English feasts'.[13] Namier, whose abrasive personality had antagonised many colleagues, was disgusted at the way his and his chief's efforts were received, and when an aggrieved Weizmann resigned his post at a Zionist Congress at Basle in 1931, Namier followed him. Although Weizmann returned to the Organisation, and the leadership, Namier did not do so. However, he did continue to work unofficially alongside the elected committee throughout the 1930s and during the Second World War.

Namier's arrival in Manchester came at a critical time in his life, and was widely seen as a coup for the University. Despite his engagement with the world of international politics, he had just published in quick succession two major books, *The Structure of Politics at the Accession of George III* and *England in the Age of the American Revolution*, and was for a time the most talked-about historian in Britain. He was added to the shortlist for the vacant chair after the head of the History Department, E. F. Jacob, read a review of the second of these books by G. M. Trevelyan, which ended by lamenting the likelihood that Namier, who had never held a permanent university post, would be obliged by financial exigency to abandon historical studies. Confidential opinions were sought from leading figures in the historical profession, all of which were highly favourable, and after interview Namier, to his great surprise, was offered the chair.

By this time Namier had lived for almost a decade in London, which he always found dirty and in some respects oppressive, but which was after all the capital city of the Empire. Manchester, and cramped lodgings in Fallowfield, proved even dirtier, its industrial atmosphere almost calculated to aggravate his chronic respiratory problems. Worse still, it was a long train journey from the centre of public affairs, and from the main sources for his research, in the British Museum and the Public Record Office. The presence of a substantial Jewish middle class made little difference to Namier, and he was even less concerned with the lives of Manchester's Jewish poor, with whom he had little or no interaction.

The city did, however, hold one major compensating attraction. Namier had been a consistent contributor to the *Manchester Guardian* since the early 1920s, at first sending reports from Central Europe to the *Manchester Guardian Commercial*, and then, once he had returned to England, as a book reviewer. The paper suited his political opinions, which at this stage of his life were still anchored on the left, although he would later shift further and further to the right. On taking up the modern history chair he approached the *Guardian* with an offer to write for them 'on the foreign side', and although this was not taken up he did forge a fruitful alliance with W. P. Crozier, editor from 1932 onwards, resulting in more book reviews and occasional articles.[14] Crozier's sympathy for Zionism, and hostility to the Nazi regime in Germany following Hitler's assumption of power, made him a kindred spirit as far as Namier was concerned. On Crozier's death in 1944, Namier wrote that 'the Jews and especially the Zionists – in fact all people who suffer and strive for a human cause – have lost a sensitive friend and an effective champion'.[15]

Disillusionment with Zionist politics had not made Namier any less a Zionist. As well as working informally through his political contacts to assist the Zionist Organisation, he wrote endless magazine articles and letters to newspapers to protest injustice and assist the Zionist cause. He was also tireless in his work to raise funds for Jewish charities and to assist refugee academics to find work in Britain through the London-based Academic Assistance Council. In this enterprise he found the University authorities in Manchester highly sympathetic, and he acted as liaison between the Council and the vice-chancellor's committee. His contribution to Jewish life in Manchester was, however, limited by the amount of time he continued to spend in London and compromised by the angularity of his personality. He did what he could for undergraduate Zionist clubs but seems to have kept his distance from the Jewish groups in the city at large, which in turn regarded him with an increasingly cold eye. Even his unilateral attempts to organise an appeal for funds for a memorial to Samuel Alexander fell foul of prominent Jews in the city who felt, not unreasonably, that they too had a right to be involved.[16] More and more he regarded London as his base, maintaining a *pièd-a-terre* in or near Manchester but only coming up for a few days each week in term time. Though devoted to his students, especially the bright ones keen on research, he hankered more and more after a chair in Oxford or Cambridge, instead of 'this town, in which I somehow never manage to settle down', and eventually retired at the first opportunity in 1953 to join the recently established History of Parliament project in London.[17]

Conclusion

While Alexander seems always to have found life in Manchester attractive and fulfilling, and was deeply respected by the Jewish communities in the city (as evidenced by their resentment at being excluded from schemes to honour his memory), the same could scarcely be said of Weizmann and Namier. Weizmann disliked Manchester at first and had little time for its working-class Jews, though he did eventually make friends among the middle classes, who were such a mainstay of the Zionist movement nationally and internationally. Namier always disliked the physical environment of the city, and, aside from his writing for the *Manchester Guardian*, always looked to London for political and academic stimulation.

Both Weizmann and Namier were committed to Zionism before they arrived in Manchester. In Weizmann's case the decisive factor in pushing him towards belief in a Zionist solution to the 'Jewish question' was undoubtedly his upbringing and immediate experience of the brutal realities of life in the *shtetls* of late nineteenth- and early twentieth-century Russia. This worked upon a passionate temperament which naturally induced devotion to an ideal. Namier's childhood in Galicia was different, his family's history and circumstances robbing him of any obvious roots. He sought refuge first in an idealisation of the British Empire, and then, when he had finally come to regard himself unequivocally as a Jew in response to what he observed of the antisemitism of Eastern and Central European nationalists and the weakness of the British establishment, in the necessity for Jews to establish a nation-state of their own. Zionism was a corollary of this personal redefinition. Alexander, on the other hand, was raised far from the Pale of Settlement, and never frustrated in academic ambitions by his Jewishness. This did not prevent him from sympathising intensely with those whom Namier would have called his 'co-racials', a sympathy that eventually, in the 1930s, pushed him closer to an acceptance of the Zionist imperative. But he did not have the same inner demons to appease and he remained, essentially, an armchair Zionist.

Notes

1 'Manchester and Zionism', *Manchester Guardian*, 24 May 1951, p. 2. I owe this reference to Stuart Jones.
2 Biographies of Alexander can be found in *ODNB* and *Australian Dictionary of Biography*, https://adb.anu.edu.au/.
3 Norman Rose, *Chaim Weizmann: A Biography* (London: Weidenfeld and Nicolson, 1986), p. 173.
4 JRL, ALEX/A/1/1/303, Weizmann to Alexander, 3 February 1922.
5 L. B. Namier, 'An appreciation', *Manchester Guardian*, 4 September 1938, p. 5.
6 JRL, ALEX/B/4/32/3, Namier to John Laird, 18 November 1938.
7 See, in general, Rose, *Weizmann*.
8 Ibid., p. 89.
9 See, in general, Julia Namier, *Lewis Namier: A Biography* (London: Oxford University Press, 1971); Norman Rose, *Lewis Namier and Zionism* (Oxford: Clarendon Press, 1980); D. W. Hayton, *Conservative Revolutionary: The Lives of Lewis Namier* (Manchester: Manchester University Press, 2019).
10 Norman Rose (ed.), *Baffy: The Diaries of Blanche Dugdale 1936–1947* (London: Valentine, Mitchell, 1973), p. 124.
11 Hayton, *Conservative Revolutionary*, p. 31.
12 Rose, *Lewis Namier*, p. 32.
13 Rose, *Weizmann*, p. 290.
14 JRL, GDN/A/N/2/13, E.T. Scott to James Bone, 5 May 1931.
15 L. B. Namier, 'An appreciation', *Manchester Guardian*, 18 April 1944, p. 5.
16 JRL, ALEX/C/4/7, N. Laski to J.S.B. Stopford, 9, 13 September 1938.
17 History of Parliament Trust archive, London, N–52, Namier to Lady Sandwich, 14 February 1934.

9

Shared friendship and divided politics:
Patrick Blackett and Michael Polanyi
in Manchester

Mary Jo Nye

Patrick Blackett (1897–1974) arrived at the University of Manchester in 1937 as Langworthy Professor of Physics and head of department, succeeding William Lawrence Bragg. Blackett had become internationally known for his experimental work in nuclear physics during ten years in Ernest Rutherford's Cavendish Laboratory (1922–33) in Cambridge and four years as Professor of Physics at Birkbeck College in London. Arriving in Manchester, where he previously had served as an external examiner, Blackett became close friends with the distinguished Hungarian and Jewish-born Michael Polanyi (1891–1976), who had come to Manchester in the autumn of 1933. Polanyi had earlier declined an offer at Manchester and decided to stay at Berlin's Kaiser Wilhelm Institute for Physical Chemistry, headed by Fritz Haber. The appointment of Adolf Hitler as German Chancellor and the promulgation of antisemitic laws that purged Jews from the German civil service necessitated Polanyi's departure from Germany, and he was fortunate in receiving another offer from Manchester as head of the University's Physical Chemistry Department and Laboratory.[1]

Scientific work and personal friendship

Polanyi and Blackett each brought major scientific reputations and accomplishments to Manchester. While in Berlin, Polanyi had become an international leader in the field of reaction kinetics, studying chemical reaction rates in gases and beginning the development of a theory of the transition state with the American physical chemist Henry Eyring during Eyring's brief stay in Berlin. Among Polanyi's colleagues in Berlin, Juro Horiuchi, who was studying hydrogen exchange and catalytic reactions of hydrogen, came with Polanyi to Manchester, as did Andreas Szabo, who was working on hydrogen and halide substitution in organic chemical reactions. Meredith G. Evans became Polanyi's principal collaborator at Manchester in the field of reaction kinetics and the transition state, along with G. N. Burkhardt and postdoctoral fellow Richard Ogg. Melvin Calvin, the 1961 Nobel Laureate in Chemistry, arrived for two years of postdoctoral study in 1933 and later traced the origins of his discovery of the mechanism of photosynthesis in chlorophyll to research with Polanyi on metalloporphyrins, such as chlorophyll, and their electronic behaviour.[2]

Blackett's main field of research when he arrived in Manchester in 1937 was the study of subatomic particles in cosmic rays. While at the Cavendish, he had used the cloud chamber to study subatomic particles and famously published a paper in 1925 with photographic tracks of a proton ejected from a nitrogen nucleus when an alpha particle was captured by the nitrogen nucleus. A later paper in 1933 with Giuseppe Occhialini included a stunning display of photographs showing the track of a positive electron in a shower of cosmic rays and proposed an explanation of the particle's identity using Paul Dirac's new theoretical prediction of the existence of an anti-electron, which became known as the positron. The move to Birkbeck College gave Blackett a professorship and control of his own laboratory, and the subsequent move to Manchester gave him better facilities, as well as a prestigious position previously held by Arthur Schuster, Rutherford and Bragg. He built a research group in cosmic-ray physics that included John G. Wilson, George D. Rochester, Leon Jánossy, Clifford C. Butler and Bernard Lovell. By the late 1930s Blackett's research on very penetrating particles in cosmic rays contributed to the discovery of the mu-meson or muon, the subject of a 1938 conference in Manchester where he hosted friends Werner Heisenberg, Bruno Rossi and Homi Bhabha, among others.[3]

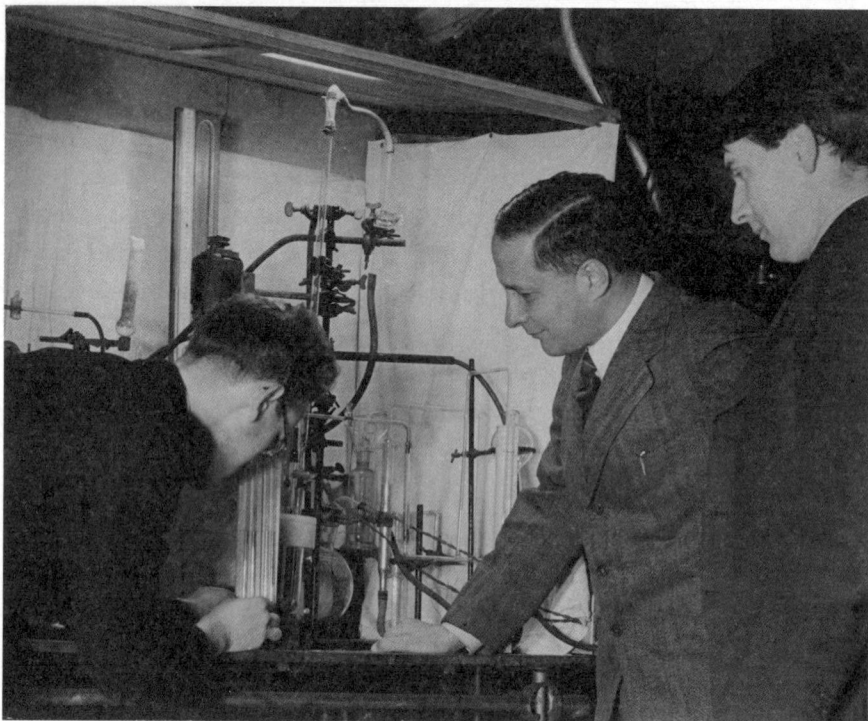

9.1 A British Council photograph of Michael Polanyi and his colleague Dr A. G. Evans watching an experiment in his laboratory. Courtesy of the University of Manchester.

By that time Blackett and Polanyi had become close friends. They, and their wives Costanza Blackett and Magda Polanyi, regularly saw each other. Blackett addressed Polanyi by the familiar diminutive 'Mischi', and Polanyi signed his letters to Blackett 'Misi'. While renting a holiday cottage in Wales, the Blacketts arranged for the Polanyis to visit and stay nearby. During and just after the war, when the Blacketts lived in London, where Blackett was a leader of operational research at the Admiralty, the 'two Pats', Patrick and Costanza (nicknamed 'Pat'), stayed with the Polanyis when in Manchester. They all kept in touch after Blackett had taken a professorship at Imperial College in 1953 and Polanyi had become a fellow at Merton College, Oxford, in 1959. Their talk now was of their honours and their lecture tours and their experiences of growing older.[4]

Political trajectories

All through these years, Polanyi and Blackett frequently argued over political matters, and they were sometimes frustrated with each other. In a letter in October 1941 Polanyi expressed chagrin at what he saw as a hostile tone in Blackett's conversation with him earlier in the day, noting, 'We have always disagreed, yet maintained an entirely genuine link of sympathy.' Blackett responded by denying any personal hostility, but admitted to being hostile to 'some of your views, and as you are to mine'.[5] Nor were these controversies confined to the dinner table or staff rooms and the large tables of the Manchester cafeteria. For well over thirty years, Polanyi and Blackett participated in heated debates that took place in British newspapers and popular periodicals, on BBC radio programmes, as well as in scientific organisations and on University and governmental committees. The two men were scientists who assumed public roles, and their views reflected a major divide among British scientists at the time on matters having to do with socialism and capitalism, developments in the Soviet Union, Britain's national economic organisation, the social responsibility of the scientist, and the autonomy of the scientific community.

This is not to say that British or Manchester scientists typically engaged in political discussions. By way of contrast, consider a 1969 interview by Edinburgh sociologist David Edge with William Lawrence Bragg, then 79 years old. When asked about his political views and about those of his father, William Henry Bragg, Sir Lawrence replied that they were both apolitical, and that he thought they generally voted Conservative, but that he saw no relationship whatsoever between scientific work, scientific values and politics. Blackett and Polanyi provide a strong counterpoint to Bragg's view.[6]

At first glance, the family backgrounds of Polanyi and Blackett might suggest different political dispositions than what became Polanyi's laissez-faire liberalism and Blackett's Labour Party socialism.

Born in Budapest to non-observant Jewish parents, Polanyi was the son of a civil engineer and railway entrepreneur who died in 1905. His widowed mother, originally from Vilnius, which was part of Russia at the turn of the century, regularly entertained a salon of intelligentsia in Budapest whose members were sympathetic with Bolshevism and socialism, as were Polanyi's two sisters and two brothers. His brother Karl became a well-known economic historian who described the inevitably socially divisive nature of the market economy.[7] In contrast to his mother and siblings, Michael even as a young man was noted for his scepticism about socialist ideologies, and it is possible to compare his later friendly differences with Blackett to his lifelong disputes with his family, especially Karl. Polanyi completed a medical degree in Budapest in 1913 and began studies in physical chemistry in Karlsruhe before serving in the Austrian army as a military surgeon and finishing his doctoral thesis in physical chemistry in Budapest while on sick leave. He served in the short-lived Hungarian Republic's Ministry of Health until a revolutionary communist government was installed in the spring of 1919, which was followed by a coup d'état establishing a new Christian and National government in Hungary. Polanyi was dismissed from a position at the University of Budapest and saw no future for himself in an increasingly antisemitic Hungary, despite his baptism in the Catholic Church in 1919. He arrived in Berlin in the autumn of 1920.[8]

Blackett was a Londoner by birth, the only son and the second of three children of a stockbroker and the daughter of an army officer. His elder sister Winifred practised for a while as an architect, and his younger sister Marion became the psychoanalyst Marion Milner. As a 13-year-old boy, Blackett entered Osborne Royal Naval College and moved on to Dartmouth Royal Naval College. When war broke out in 1914, Blackett was sent immediately into action, and participated in the Battle of the Falkland Islands in the South Atlantic in 1914 and the Battle of Jutland in 1916. After the Admiralty sent Blackett and other young officers to Cambridge in 1919, he resigned from the navy and embarked on a career in physics. At Magdalene College in Cambridge, Blackett became friends with

Kingsley Martin, later the editor of the left-wing *New Statesman*. Blackett had voted Conservative in 1918, but he could be found campaigning for Hugh Dalton and voting Labour in the 1922 General Election. During the General Strike in 1926, having already published some of his first major scientific papers, Blackett ferried copies of the *British Worker* by car from London to Cambridge. Blackett now was firmly committed to the left.[9]

Two conceptions of science and society

Polanyi became aware of Blackett's political views and activities well before Blackett joined the Science Faculty in Manchester. When Polanyi arrived in Manchester in the autumn of 1933, the National Government was in office: an uneasy Conservative-dominated coalition under former Labour leader Ramsay MacDonald, and with some Liberal support. There was much talk of the need for central planning, a development that was anathema to Polanyi, who had joined a seminar in Berlin in 1928 that was studying Soviet economic planning. Polanyi organised his own study group on economic systems in 1930. He visited the Soviet Union for the first time in 1928, at the invitation of physical chemist Abraham Joffe. Further trips in 1930, 1932 and 1935 acquainted him with Five-Year Plans in the Soviet Union, and he knew that Joffe had tried to protect fundamental research from the ideological and bureaucratic interference embodied in planning. Polanyi's first economic publication appeared in November 1935 in the *Manchester School of Economic and Social Studies*, not long after he had become friends in Manchester with the economist John Jewkes. Polanyi's article was a fact-based and seething indictment of the failures of Soviet planning in industrial and agricultural production. He further warned of the dangers of central planning to fundamental scientific research, which had to be free from deadlines and from demands that science should meet social needs.[10]

The notion that social and economic needs inspired scientific research and progress had become a matter of great interest to Blackett after he encountered the Russian scientist Boris Hessen's paper on 'The social and economic roots of Newton's *Principia*' in 1931. Blackett later said that he had been disappointed by the usual histories of science that treated science and scientists as if they were 'living so to speak in a social vacuum'. In 1934, after having moved to Birkbeck College, Blackett began speaking on BBC radio in a series of programmes on science and society organised by left-wing scientists Julian Huxley, Hyman Levy and John Desmond Bernal. Blackett argued that scientists cannot be aloof from politics because they are dependent on material and moral support from government and industry. In Blackett's view, it behoves scientists to make clear to the public the virtues of science, rather than risk the development of anti-scientific and anti-intellectual movements. Turning to MacDonald's policies, Blackett spoke out against planned capitalism as a 'third way' between laissez-faire capitalism and Soviet-style communism. That is not the way to go, Blackett argued, because it leads to fascism. What is needed is 'complete Socialism. Socialism will want all the science it can get to produce the greatest possible wealth.'[11]

In contrast to Blackett, Polanyi was developing plans in the mid-1930s for an educational film that would teach basic components of economics as a kind of conveyor belt of production and consumption. His conception, which made use of cartoon-like diagrams, made its debut in 1938 as a silent film and appeared in revised form in 1940 with a soundtrack. The economic system that the film taught was a quasi-Keynesian one that favoured full employment and a greater sense of communal responsibility within a free market system. It allowed some government oversight, but no central planning or social welfare measures. Polanyi's proposal was thus meant to be yet another 'third way' between laissez-faire capitalism and centralised socialism. He arranged showings of the film, which was produced by Gaumont Instructional Films, in Manchester, London, New York and elsewhere, including the Walter Lippmann Colloquium in Paris, a seminar of intellectuals engaged in the construction of a renewed liberalism. He gave copies of the film to the Workers' Educational Association and intended the revised version, 'Unemployment and Money', to be taught in schools and colleges. In addition, as a more permanent contribution to economics, he published the book *Full Employment and Free Trade* in 1945.[12] In 2012 the scholar Eduardo Beira made the film more accessible by digitising it.[13]

If economics was one point of discussion and dissension between Blackett and Polanyi, a more troubling one was disagreement over the Soviet 'experiment' and the claims that it had produced benefits in the form of funding for science and its applications. Blackett had visited Moscow for conferences in September 1935 and September 1937, and Polanyi's last visit had been in 1935 when he saw his sister Laura's daughter Eva Striker in Moscow. She was arrested in May 1936 on false charges of plotting against Stalin and was imprisoned in solitary confinement in Lubianka. Her estranged husband, the Viennese physicist Alex Weissberg, who was working at the prestigious Physical Technical Institute in Kharkov (Kharkiv), tried to gain her release, as did other family members, influential scientists and her old friend Arthur Koestler, who later said that her story was the inspiration for his novel *Darkness at Noon* detailing Stalinist terror and purges. In March 1937 friends were trying to get Weissberg released after his arrest on charges of scheming to kill Stalin. Eva was finally allowed to return to Vienna in September 1937, but Weissberg was handed over to the Gestapo following the Stalin–Hitler pact of August 1939, along with other Austrian, German and Hungarian prisoners. Another of the prisoners was Blackett's former assistant Fritz Houtermans, whose release Blackett had tried to secure, as he discussed with Polanyi. The Gestapo freed Houtermans in Berlin, but Weissberg spent most of the war in prisons and camps.[14]

Blackett was in Wales when the Stalin–Hitler pact was announced, and he wrote to Polanyi that he would not be surprised if Polanyi felt some 'Schadenfreude' over the Russians' behaviour. Blackett added that he was completely surprised, as well as dismayed, as Polanyi would expect.[15] It now became clear that there would be war immediately in Europe. Blackett was already serving on the committee organised by Henry Tizard in 1935 to advise the Air Ministry on priorities for national defence. After the committee recommended the development of radar, against the objections of

Winston Churchill's scientific advisor Frederick A. Lindemann, it was dissolved. Its successor was the Maud Committee, constituted in the spring of 1940, which advised the government on the feasibility of Britain's developing a fission bomb. Blackett was the lone committee member who recommended against an independent British effort and urged joint collaboration with the United States, the path that ultimately prevailed. After serving in several different capacities as advisor on defence, Blackett became director of Naval Operational Research in the Admiralty in 1942, where he remained until the war's end.[16]

Immediately following the Stalin–Hitler pact, Polanyi asked Blackett to help him get defence work, but that proved impossible even after Polanyi became a naturalised British citizen in September 1939. Thwarted in his efforts to get involved with war work, Polanyi turned to further economic and political writing, responding directly in *The Contempt of Freedom: The Russian Experiment and After* (1940) to Bernal's 1939 book *Social Function of Science*, in which Bernal praised Soviet support of science and advocated central funding for science and industry in the UK. Polanyi began meeting in March 1941 with a group called the Society for Freedom in Science, which included Oxford scientists John Baker and Arthur George Tansley, and which aimed to fight any postwar attempt to allocate funds by means of central planning. In response to a circular that solicited new members, the German refugee physicist Max Born, who was in Edinburgh, wrote to Blackett that he would not join the Society in its attack on socialist scientists whom Baker had called 'gangsters' in a review of Bernal's book. Blackett reacted to Polanyi's 1940 book by writing that he was dismayed not only by its attitude towards the Soviet Union, but 'to all that is generally called progressive politics—"progressive obsessions"—in your words'. The Russians again became allies of the UK following Germany's attack on the Soviet Union in June 1941, and Blackett told Polanyi that he admired what he thought was a remarkable military feat in the Russian resistance against Germany during the previous four months.[17]

As for postwar central planning of British science, Blackett became president of the left-wing Association of Scientific Workers from

1943 to 1947, which argued against centralisation of government scientific offices into a single Ministry of Science. He was a member of the Royal Society's postwar committee that in 1946 supported, as Polanyi recommended in a letter to the committee, the 'natural' development of science by the most distinguished leaders. As for tying scientific research to immediate social needs, Blackett's view was that Marxist thinking could go too far in reacting against the science-for-science's sake attitude.[18]

Scientists, freedom and the Cold War

Blackett was now in Manchester at work on cosmic rays and then a new theory of the Earth's magnetic field, while also serving as dean of his faculty and pro vice-chancellor of the University. In the early 1950s he built a research group on rock magnetism that included Edward (Ted) Irving, S. Keith Runcorn and John Clegg, and fostered revival of the hypothesis of continental drift. Polanyi, on the other hand, resigned his chair in Physical Chemistry in 1948 in exchange for a professorship created for him in the Faculty of Economics and Social Studies. The position was arranged by his friend John Jewkes and the vice-chancellor Sir John Stopford, approved by the Senate, and supported financially by the fund set up by Sir Ernest Simon in 1944 to promote the social sciences. Polanyi had increasingly been pursuing economic, social and philosophical analyses, outside physical chemistry, and this new arrangement enabled him to write the Gifford Lectures that would become the basis for his philosophical and moral critique of positivist science in *Personal Knowledge: Towards a Post-Critical Philosophy* (1958). Polanyi's *Logic of Liberty*, published in 1951, used his powerful metaphor of 'the Republic of Science' to argue for autonomy and self-governance among scientists.[19]

Blackett also was writing a powerful and controversially political book, published in the UK in 1948 as *Military and Political Consequences of Atomic Energy*, followed by an American edition of 1949 under the title *Fear, War, and the Bomb*. He revived arguments made during his wartime service on government and military committees against the strategy of winning wars from the air by saturation bombing of civilian centres, as the Allies had done against Cologne and Dresden with conventional weapons, and the Americans had done against Hiroshima and Nagasaki with atomic weapons. Mass destruction could not and should not be a normal operation of war, Blackett argued, on the grounds that it is militarily ineffective and morally reprehensible. Convinced of the need for negotiations on both conventional and atomic weapons, Blackett advocated cooperation with the Soviet Union in an attitude of hopefulness about Russia and about the possibilities for peaceful uses of atomic energy. He also displayed a clear distrust of the United States and its ambitions, and he argued against British development of atomic weapons.[20]

What was Polanyi's view on these matters? He might have adopted the stance of his friend, the University of Chicago sociologist Edward Shils, who characterised Blackett's book as a Stalinist apologia. It seems that Polanyi did not agree with Shils, however. In 1946 Polanyi returned from a trip to the United States during the autumn election campaign, 'shaken to the core', as Blackett wrote to his friend Nevil Mott, by what he perceived as an almost cavalier attitude in the American press towards the use of atomic weapons. In March 1947 Polanyi wrote a memorandum in support of the renunciation of weapons of mass destruction, both atomic and biological, and advocated international cooperation on peaceful atomic energy. He further proposed that the United States should destroy its existing atomic bombs. Polanyi did not participate, however, in the Pugwash conferences on disarmament after the organisation's founding in Nova Scotia in 1957, while Blackett did participate and knew one of the founders, Joseph Rotblat. Michael Polanyi's younger son John, a theoretical chemist and 1986 Nobel Laureate, became a leading figure in the Pugwash movement.[21]

Another shared experience for Polanyi and Blackett was scrutiny by government intelligence agencies, each of them falling under suspicion for possible communist ties. It is ironic that this happened to Polanyi despite the fact that he was a well-known and well-published anti-communist. Polanyi's difficulties lay not in the UK but in the United States, where he gave a series of anti-Soviet and anti-authoritarian lectures on the 'Logic of Liberty' in the spring of 1950 at the University of Chicago. He soon applied to the US State Department for a non-quota immigrant visa that would enable him to accept the University of Chicago's offer of a chair in social philosophy to begin in the autumn of 1951. Despite letters from Edward Shils and other prominent Chicago academics, as well as interviews with a US consular official in Liverpool, the visa was not forthcoming. Initially Polanyi thought that the problem might be his long-ago involvement with leftist student societies in Budapest, but it later became evident that he was being barred because his name was associated with the Institute for Free German Culture, where he gave a lecture in 1942 without knowing that it was a pro-Soviet organisation. His name was also listed during 1946–47 as a sponsor by the Society for Cultural Relations with the Soviet Union, which enabled the translation of Russian documents into English, many of them scientific articles. In 1953, when a visa seemed in sight, Polanyi was again interrogated, this time on the political opinions of his brother Karl, whose wife was a former member of the Hungarian Communist Party. He decided not to pursue the matter further.[22]

Over a longer period, Blackett was under scrutiny both in the UK and the US. When he and his wife Costanza made a refuelling stop in Tampa, Florida, on a flight to Toronto following a cosmic-ray conference in Guanajuato, Mexico, in 1951, they were detained and questioned about his political views. The FBI kept files on him, which were dotted with conspicuous erroneous claims and names. Despite his status as the 1948 Nobel Laureate in Physics, Blackett remained under considerable criticism until the 1960s both in the US and the UK from scientists, public intellectuals and the press, for his advocacy of cooperation with the Soviet Union and his opposition to the further development of atomic weapons.[23]

In May 1949 Blackett's name appeared on novelist George Orwell's blacklist of 38 crypto-communists or fellow travellers that Orwell sent to a friend in the British Foreign Office. Suspicion was rife generally against leftist scientists in this period. Theoretical physicist Allan Nunn May, who had worked in Canada for the Manhattan

9.2 Patrick Blackett with his daughter Giovanna and his wife Costanza in 1948 in Manchester, on receiving the news of the award of the Nobel Prize in Physics. Courtesy of Kate Bloor.

Project, had been sentenced to ten years in prison after his arrest in 1946 for giving information during the war and afterwards to the USSR. Klaus Fuchs, who was head of the theoretical division at Harwell and working with John Cockcroft, was arrested in February 1950 for passing information. The cosmic-ray physicist Bruno Pontecorvo, who was also working at Harwell, defected to the Soviet Union in August 1950.[24] By the 1950s the British security services in MI5 had been keeping files on Blackett for many years, dating back to 1931. The official final summary for these files says that he was assessed as having an intellectual interest in communism and that he was close to the Communist Party in the 1940s but not a member. By 1952 the security services concluded that the only risk that Blackett posed was that he might inadvertently mention valuable information to questionable associates. There was no issue regarding his personal integrity.[25] Still, as late as 1960 Blackett had to obtain a special visa at the American embassy in London before he could enter the US.[26]

By 1953 Blackett was living in London after becoming head of Physics at Imperial College, where the Blackett Laboratory still carries his name. He was president of the Royal Society from 1965 until 1970, and was made Lord Blackett of Chelsea in 1969, despite his earlier misgivings about the politics of the House of Lords. Unlike Polanyi, Blackett had never taken a PhD, and he quipped to his friend Edward (Teddy) Bullard that at least he had remained Mr Blackett until he retired.[27] Polanyi retired from Manchester in 1958 and moved to Oxford in 1959 after election as a senior research fellow at Merton College. His book *Personal Knowledge* had appeared in 1958, and he was becoming increasingly well known in the philosophy of science for his ideas on scientific apprenticeship, tacit knowledge and scientists' resistance to change in their disciplinary traditions. It became standard to discuss Polanyi's philosophy of science in relation to the account of the power of scientific paradigms set out in Thomas Kuhn's 1962 book *The Structure of Scientific Revolutions* and Karl Popper's criterion of falsifiability as the definition of a scientific theory in his 1959 book *The Logic of Scientific Discovery*.

In the 1950s and 1960s Polanyi became increasingly focused on the philosophy of science and the philosophy of knowledge, spending longer periods on lecture tours and professional visits in the United States. In addition to his administrative and research work at Imperial College, Blackett concerned himself with the future of atomic energy and atomic weapons, as well as scientific development in under-developed countries, serving as scientific advisor to Harold Wilson's Labour government and to Jawaharlal Nehru's new Indian government. With the horrors of the 1930s and the Second World War at an end and the global confrontation between the US and the USSR a new concern, differences between Polanyi and Blackett were less sharp, and in some political matters there was rapprochement.

The deep friendship between Magda Polanyi and Costanza Blackett played a role in the cementing of the relationship between the two men. They all appear to have seen each other less frequently, however, after leaving Manchester. The long friendship remained an affectionate one, as expressed in letters from Magda to Costanza in the 1960s, writing that they must get together soon.[28] Thus, Polanyi and Blackett were bound together personally by shared conviviality and sympathy, along with their commitment to scientific values and to scientific progress. Their differing political views mirror the opinions and choices that separated and bound together many scientists and public intellectuals of their era. They remained friends, even if they were sometimes friends at odds.

Notes

1 The principal biographical sources for Patrick M. S. Blackett are Sir Bernard Lovell, 'Patrick Maynard Stuart Blackett, Baron Blackett, of Chelsea', *Biographical Memoirs of Fellows of the Royal Society*, 21 (1975), 1–115, and Mary Jo Nye, *Blackett: Physics, War, and Politics in the Twentieth Century* (Cambridge, MA: Harvard University Press, 2004). For Michael Polanyi, see Eugene P. Wigner and Robin A. Hoskin, 'Michael Polanyi, 12 March 1891–22 February 1976', *Biographical Memoirs of Fellows of the Royal Society*, 23 (1977), 413–48; William T. Scott and Martin X. Moleski, SJ, *Michael Polanyi: Scientist and Philosopher* (Oxford: Oxford University Press, 2005); and Mary Jo Nye, *Michael Polanyi and His Generation: Origins of the Social Construction of Science* (Chicago: University of Chicago Press, 2011). This chapter draws heavily on Nye, *Blackett*, and Nye, *Michael Polanyi*. It is also a revision of my article 'Manchester friends at odds: Michael Polanyi, P. M. S. Blackett and the scientist as political speaker', *Manchester Region History Review*, 18 (2007), 40–63. I thank *MRHR* for permission to republish parts of that article and I am grateful to Stuart Jones for comments and suggestions on this revision.
2 Nye, *Michael Polanyi*, pp. 132–7; Scott and Moleski, *Michael Polanyi*, p. 157.
3 Nye, *Blackett*, pp. 43–62.
4 Nye, 'Manchester friends', 123 nn. 5–9.
5 Nye, 'Manchester friends', 108.
6 David Edge's interview with Sir Lawrence Bragg, 20 June 1969. American Institute of Physics, https://www.aip.org/history-programs/niels-bohr-library/oral-histories/28531 (accessed 14 September 2023).

7 Karl Polanyi, *The Great Transformation* (New York: Farrar and Rinehart, 1944).

8 Nye, *Michael Polanyi*, pp. 3–17.

9 Nye, *Blackett*, pp. 15–28.

10 Nye, 'Manchester friends', 113–14; *Michael Polanyi*, pp. 153–8.

11 Nye, *Blackett*, pp. 32–3.

12 Nye, *Michael Polanyi*, pp. 158–69.

13 Gabor Biro, *The Economic Thought of Michael Polanyi* (New York: Routledge, 2020); Mary Jo Nye, 'Transforming economics with a film projector', *Metascience*, 29 (2020), 139–42.

14 Nye, *Michael Polanyi*, pp. 198–200, 339 n. 67.

15 Nye, *Blackett*, p. 38.

16 Nye, 'Manchester friends', 114–15.

17 Ibid., 115–16.

18 Nye, *Blackett*, pp. 39–40.

19 Ibid., pp. 120–32; Nye, *Michael Polanyi*, p. 140; Scott and Moleski, *Michael Polanyi*, p. 204. For an analysis of Polanyi's arguments, see 'Foreword' to Michael Polanyi, *Personal Knowledge: Towards a Post-Critical Philosophy*, enlarged edition with a new foreword by Mary Jo Nye (Chicago: University of Chicago Press, 2015), pp. xi–xxv.

20 Nye, 'Manchester friends', 117–18.

21 Ibid., 118–19.

22 Nye, *Michael Polanyi*, pp. 27–9.

23 Nye, *Blackett*, pp. 92–3.

24 Ibid., p. 92.

25 The National Archives, KV2 The Security Service: Personal (PF Scrics) Files: KV2/3217, 3218, 3219, 3220 online catalogue at discovery/nationalarchives. gov.uk. I thank Michael Holzman for alerting me to the availability of these records.

26 Nye, *Blackett*, p. 92.

27 Edward Bullard, 'Patrick Blackett … an appreciation by Sir Edward Bullard', *Nature*, 250 (1974), 370.

28 Nye, 'Manchester friends', 123 n. 7.

Formed in Europe, shaped in Manchester: refugee engineer Professor Franz Koenigsberger

Jonathan Aylen

The appointment of Franz Koenigsberger as a lecturer in Mechanical Engineering at the Manchester Municipal College of Technology in 1947 seems unexceptional. The college, which was also the University's Faculty of Technology, was enthusiastic about linking practical developments in machine tools with academic research.[1] Koenigsberger was a pioneer in the use of fabricated steelwork to replace heavy castings in machine tools. He had published a prize-winning technical paper, 'The Application of Fabricated Construction to Machine Design', in 1945 and given a popular wartime lecture course on 'Theory and Design of Welded Structures' at the College of Technology.[2] He became the UK's first ever Professor of Machine Tool Engineering at the Manchester College of Science and Technology (later UMIST) in July 1961 and wrote three major books on machine tool design.[3] The simple facts of his appointment, however, conceal Koenigsberger's peripatetic pre-war exile as a Jewish refugee from Nazi Germany and his escape in 1938 to the safe haven of the city of Manchester, where he quickly became head of engineering at a local machine tool firm.[4]

Franz Koenigsberger was born in Germany in November 1907 and completed his education with a diploma in Engineering from the Technische Hochschule at Berlin-Charlottenburg, graduating in 1931. He remained at the Technische Hochschule in the machine tool laboratory working for Dr Georg Schlesinger, a fellow Jewish

V3.1 The Renold Building, opened in 1962, the year after Franz Koenigsberger's appointment as Professor of Machine Tool Engineering at the Manchester College of Science and Technology (later UMIST). It was the centrepiece of the 1960s campus development.

engineer. In his words, 'he married the boss's daughter', Lilli Schlesinger, in 1934. But the Nazi law of April 1933 for 'Restoration of a Professional Civil Service' – colloquially known as the *Berufsbeamtengesetz* – excluded Jews from public employment. The Hochschule was a public university and Jewish academics were forced out.

Koenigsberger found employment locally as a senior draughtsman at G. Kärger in Berlin, a manufacturer of high-precision tool-room lathes. Koenigsberger was reassured that he had nothing to fear from anti-Jewish laws. The owner's defence of his Jewish employees was to lead to his premature death in a concentration camp before the war.

In the interests of self-preservation, Koenigsberger moved with his wife to Liège in Belgium, and then on to a senior role at Ansaldo, a large engineering firm in Genoa in April 1936. Within a couple of years, Mussolini's regime adopted antisemitic laws similar to those in Germany. Koenigsberger was again forced to move on, back to Belgium in August 1938. After a stressful wait, he gained a crucial visa to work in Manchester in October 1938.

Once in Manchester, Koenigsberger moved jobs to become chief designer at the new firm of Cooke and Ferguson in June 1939. His father-in-law also became a refugee, moving first to Belgium and then becoming head of a machine tool research scheme in Loughborough at the end of 1938.[5] All told, some 800–1,000 German Jewish engineers found safe haven in Britain during the 1930s.[6] Assimilation was helped by the common language of engineering drawings and the fact that Britain's rearmament was strongly supported by imported German equipment.

Franz Koenigsberger was by no means the only refugee to survive and prosper through employment at Manchester University. Michael Polanyi, a leading Hungarian-German chemist, took a stand against the *Berufsbeamtengesetz* and was controversially offered a chair at Manchester (see Chapter 9).[7] A group of concerned academics at Manchester responded to the persecution of German and other European academics by setting up an ad hoc rescue committee in May 1933, led by the vice-chancellor, with the active support of senior staff such as the historian Lewis Namier and the physicist Lawrence Bragg. This gained official support from the University Senate and developed through cooperation with the UK-wide Academic Assistance Committee.

The committee acted as a clearing house for distressed academics, offering temporary research appointments funded by charitable donations to support refugees while they found permanent university roles. Manchester's welcome to refugees was consistent with a long-established policy of recruiting the best possible academics.

Among the 33 academics who reached the University from Germany, Italy and Spain was Dr Rudolf Peierls in 1933, who was later to move to Birmingham and write the crucial memorandum with Otto Frisch showing that an atomic bomb was a practical possibility.

*   *   *

Advice from Lucy Bonner, William Craig, David Eaton and Joe Murphy is gratefully acknowledged.

Notes

1 Franz Koenigsberger, 'Machine tool engineering at UMIST', in D. S. L. Cardwell (ed.), *Artisan to Graduate: Essays to Commemorate the Foundation in 1824 of the Manchester Mechanics' Institution, now in 1974 the University of Manchester Institute of Science and Technology* (Manchester: Manchester University Press, 1974), pp. 221–30.

2 Franz Koenigsberger, 'The application of fabricated construction to machine design', *Proceedings of the Institute of Mechanical Engineers*, 152:1 (1945), 245–74. This was awarded the Thomas Lowe Gray Prize in 1944 by the Council of the Institution of Mechanical Engineers, worth £25. His public lectures are reported in Institute of Mechanical Engineers, 'Proposal for Transference: Associate Membership to Membership, Franz Koenigsberger' (received 20 July 1945); correspondence with Lucy Bonner, Archivist, IMechE, June 2023.

3 'Chair Machine Tool Engineering (Koenigsberger)', The University of Manchester (University Archives, File: VCA/7/689, May 1961); Franz Koenigsberger, *Design for Welding in Mechanical Engineering* (London: Longmans, Green, 1948); Franz Koenigsberger, *Welding Technology* (London: Cleaver-Hume Press, 2nd edn, 1953); Franz Koenigsberger, *Berechnungen, Konstruktionsgrundlagen und Bauelemente spanender Werkzeugmaschinen* (Berlin: Springer, 1961).

4 Franz Koenigsberger, *Anvil or Hammer: Reminiscences and Thoughts of a Mechanical Engineer Who Would Never Give In* (Oxford: Rewley, 1980).

5 'Georg Schlesinger to take charge of joint research', *Aircraft Production*, 1:2 (1938), 49.

6 Wolfgang Mock, 'Engineers from Germany in exile in Britain, 1933–1945', in Werner E. Mosse (ed.), *Second Chance: Two Centuries of German-speaking Jews in the United Kingdom* (Tübingen: J.C.B. Mohr, 1991), pp. 347–60.

7 Bill Williams, *Jews and Other Foreigners: Manchester and the Rescue of the Victims of European Fascism, 1933–1940* (Manchester: Manchester University Press, 2011).

10

W. Arthur Lewis and economic development: a Manchester story

Gerardo Serra

Grounding W. Arthur Lewis

Born on the small island of St Lucia (British West Indies), William Arthur Lewis (1915–91) was the first scholar of African Caribbean descent to be awarded, in 1979, the Nobel Prize in Economics. It was one of Lewis's many 'firsts': he was the first West Indian to win a Nobel Prize in any field, the first Black professor at a British university, and the first president of the Caribbean Development Bank.[1] Some of the insights that led him to become one of the founders of development economics occurred to him in Bangkok when, in 1952, he had the sudden realisation 'that all one needed to do was to drop the assumption – then usually made by neoclassical macroeconomists – that the supply of labor was fixed'.[2] In 1957 Lewis travelled to Ghana to become the financial advisor to the first British colony in sub-Saharan Africa to become independent. In 1958 he became the first principal of African Caribbean descent of the University College of the West Indies (later the University of the West Indies, where Lewis served as the first vice-chancellor) before taking up an appointment at Princeton University in 1963. There he spent most of his time until retirement.

These biographical snippets portray Lewis as a cosmopolitan intellectual, who in different guises as an economic theorist (albeit one always inspired by history and guided by empirical concerns), as a university head and as an international 'expert' built institutions of learning, advised governments, and changed the way we think about growth and development. Yet, this chapter argues, none of these elements in Lewis's life and work would be fully legible without grounding them in a specific time and place: Manchester in the 1940s and 1950s. Manchester is a central node in the imperial topography within which Lewis, as a colonial subject animated with anti-colonial ideas, built his identity as a social scientist and as an activist. The complementarity of imperial topographies and Lewis's global reach is exemplified in his passport (figure 10.1), with stamps from Barbados, Ghana, Egypt, Japan, Mexico and Nigeria, but also Italy, Canada and the United States.

In conversation with Lord Simon of Wythenshawe (an industrialist, politician and chair of the University of Manchester's Council), Lewis pointed out that 'the world is his laboratory', and that 'if he wants to write sensible and practical things he must spend a lot of time in other countries'.[3] Yet grounding him in Manchester makes visible the critical interconnections between his lives as an economist, anti-colonial intellectual and social activist. Despite this volume's primary focus on the University of Manchester, it is only by thinking more broadly about the entanglements of the city's history with Lewis's life that the multidimensional nature of his commitment to economic development can be holistically reconstructed. Through the lens of Manchester, this chapter shows that this was not limited to the academic task of explaining why some nations are rich and others are poor (and what the latter can do to alter their fate), but also to an anti-colonial interrogation of political and economic forms, and to community building in deprived areas.

Reimagining anti-colonialism at Chorlton-on-Medlock Town Hall

Between 15 and 21 October 1945 Manchester hosted the Fifth Pan-African Congress (PAC). This brought together intellectuals and activists such as the Trinidadian George Padmore, the African American W. E. B. Du Bois, the Manchester-based Guyanese T.

10.1 Arthur Lewis's UK passport, issued in 1953. W. Arthur Lewis papers, Box 1, Folder 12, Seeley G. Mudd Manuscript Library, Princeton University.

Ras Makonnen and many African nationalists who were about to spearhead their countries' struggle for independence (among others, Kwame Nkrumah, Hastings Banda and Jomo Kenyatta). While barely attracting any contemporary press coverage (with the key exception of the *Manchester Guardian*), the PAC represents a landmark in the

10.2 Plaque commemorating the Fifth Pan-African Congress on the front of the former Chorlton-on-Medlock Town Hall building. Photograph by Gerardo Serra.

history of anti-colonialism and is currently remembered as such in Manchester's urban landscape (figure 10.2).[4]

Chorlton-on-Medlock Town Hall, where the congress took place, was decorated with 'the flags of the Republics of Haiti and Liberia' and the flag of Ethiopia.[5] The flags of the only independent Black nations evoked simultaneously a world free of colonial domination and the necessity to imagine new political communities. Indeed, despite significant differences, the many souls of the PAC converged in emphasising the necessity to fight imperialism, abolish racial discrimination and build transnational networks of solidarity.[6] These demands were grounded in a commitment to assert the self-determination of colonial people by summoning political forms that could simultaneously provide alternatives to the nation-state and overcome the economic dependency inherent in the metropolis–colony relation.[7] In the PAC, this willingness to subvert imperial political topographies resulted in calls for the creation of a West Indian and a West African federation.[8]

At the time of the Congress, Lewis was already an affirmed economist with significant experience in academia and policy advice. He dated his pan-African convictions to his childhood in the West Indies: when he was seven years of age, his father took him to a meeting of the local Marcus Garvey association.[9] Following completion of his education in St Lucia, a prestigious scholarship took him to Britain, where he obtained a Bachelor of Commerce and a PhD from the London School of Economics. There he became very involved in the Fabian Society. Following an early interest in industrial economics, his work at the Colonial Economic Advisory Committee, the Colonial Economic and Development Council and the Colonial Development Corporation contributed to shift his focus to the economic issues of the developing world and taught him some hard lessons about the nature of policymaking.[10] Although there is no archival evidence to suggest that Lewis attended the Congress, the PAC is crucial to understanding his life and thought. Lewis's anti-colonial networks in London included several of the Congress's key individuals such as Padmore and Nkrumah.[11] In 1946 Lewis contributed to organising a Fabian Colonial Bureau conference that brought together British socialists, representatives of the Colonial Office and spokespersons from West Africa and the West Indies.[12] Despite Lewis's ambivalence towards its more radical facets, inscribing Lewis within the anti-colonial moment epitomised by the PAC allows us to capture the entanglements between his politics and his economic thinking. In his own words: 'my interest in [economic development] was an offshoot of my anti-imperialism'.[13]

Lewis's first key work, the Fabian pamphlet *Labour in the West Indies*, described the 1930s workers' protests that arose from halved export prices, widespread poverty and unemployment.[14] Lewis identified a wide range of economic policies to improve Caribbean living standards, including 'increased preferential trade treatment', import substitution, industrialisation and redistribution of income and land.[15] But to overcome their predicament as a group of small islands dependent on exports of raw materials, the West Indies ultimately required a change in political structure.[16] Throughout his life, Lewis remained a vocal proponent of a West Indies Federation. This eventually materialised between 1958 and 1962, this time with the support of the Colonial Office, eager to minimise the financial costs of impending independence (although Lewis's native island of St Lucia became independent only in 1979, the same year that he won the Nobel Prize). Due to diverging visions and incompatible vested interests, the Federation collapsed. This period largely overlapped with Lewis's tenure as principal of the University College of the West Indies and vice-chancellor of the University of the West Indies, which proved to be a veritable microcosm of the political tensions and financial difficulties of federalism.[17] This experience left Lewis disillusioned, but did not end his plea for federalism and for a holistic approach to economic issues.[18]

Two decades later he maintained that 'In the end, economics is not enough. People of different races, religions, and cultures have to learn to live peacefully with each other, and to develop pluralistic and federal institutions where this is the only way.'[19] Concluding his volume on racial conflict and economic development, this passage suggests that Lewis's federalism entailed more than the achievement of economies of scale (a key concern in the case of the West Indies). Nor did Lewis's federalism only gesture at alternative supra-national orders. Rather, it underpinned his conception of pluralist democracy, which anticipated important aspects of what political scientists would later call 'consociational democracy'.[20] This was particularly evident in West Africa. While sharing African leaders' 'anti-imperialism and their goal of an Africa united in stages', Lewis interpreted federalism as a flexible platform to mitigate at the national level the disruptive legacies of colonial rule, including high economic inequalities and arbitrary boundaries that grouped together heterogeneous peoples with diverging interests and political systems.[21] In combination with proportional representation for minorities and coalition government, federalism represented an antidote to the one-party states that came to dominate the West African political landscape shortly after independence.

Lewis's federalism was very different from Padmore's vision of a British Commonwealth modelled along Soviet lines, or from Du Bois's attempts to radically alter colonial peoples' representation at the United Nations.[22] But it was grounded in many of the same concerns, working as a foil to interrogate the mutual constitution of economic and political forms, and contributing to imagine a postcolonial world of racial equality. The emphasis on federalism makes more visible the institutionalist sensitivity of Lewis's economics, and suggests that the political ramifications of his thought are not simply an appendix to his work as an economist and policy advisor.[23] They represent a necessary embodiment of his conviction that the economics that mattered for poor countries was 'not the economics of marginal utility'; rather it was 'the economics of laws, institutions, tenures, nationality, race, religion, ideology [...] which were banished [...] as disreputable topics over seventy years ago'.[24] The rediscovery of nineteenth-century insights and concerns underpinned the academic contributions for which he is mostly remembered.

Unveiling economic development at the University of Manchester

Christmas 1947 marked the beginning of Lewis's employment at the University of Manchester.[25] In the decade he spent there in the Stanley Jevons Chair of Political Economy in the Faculty of Economic and Social Studies, he played a pivotal role in establishing development economics as a distinct and legitimate academic field. Propelled in its political importance by the Cold War and the struggle for decolonisation, development economics provided the leaders of the 'new' postcolonial nations with a vocabulary to articulate aspirations of transforming their economies and societies, while also masking uneven power relations behind new international financial institutions and transnational governance.[26] Many of these concerns were expressed in terms of increasing growth rates, savings and capital accumulation. In contrast with prevailing orthodoxy, the first generation of development economists maintained that poor countries, characterised by pervasive market failures, dependence on the export of primary products and distinctive institutions, required a new kind of economic theory, tailored to their specific conditions.[27] In forging its unique intellectual identity, development economics mobilised disparate influences that included echoes of 1920s Soviet debates on industrialisation, Keynesianism, colonial policy and nineteenth-century classical political economy.[28] The latter was particularly crucial for Lewis, whose theorising was imbued with history, but who was uninterested in setting 'out a theory of the laws of evolution of society', like 'Ricardo or Marx or Toynbee'.[29] From classical political economists, Lewis did not only draw the specific assumption that the labour supply should not necessarily be fixed. Rather, he inherited the awareness that the key to unveiling the mystery of economic development lay in the conjoined analysis of capital accumulation and income distribution.[30]

These concerns found expression in Lewis's most celebrated article, 'Economic Development with Unlimited Supplies of Labour'.[31] He started from the observation that many developing countries (such as Egypt, India or Jamaica) possess a population 'so large relative to capital and natural resources, that there are large sections of the economy where the marginal productivity of labour is negligible, zero or even negative'.[32] Countries with these characteristics present the features of a 'dual economy', with a large 'subsistence sector' employing a vast mass of workers with negligible labour productivity, and a small 'capitalist' one, which uses 'reproducible capital, and pays capitalists for the use thereof'.[33] In these cases, economic development occurs by shifting workers from the subsistence sector to the 'modern' one. Over time, this absorption leads to a reduction of workers in the traditional sector without any significant loss of productivity, and a scenario in which 'the capitalist surplus will rise continuously, and annual investment will be a rising proportion of the national income'.[34] Lewis advocated the creation of an indigenous class that, by virtue of its high savings, could promote capital accumulation, and thus facilitate the structural transformation of the world's 'periphery'. Indeed, Lewis identified this duality also at the global level, with many tropical countries stuck in relations of dependency and adversarial terms of trade for their main (typically agricultural) exports.[35] But while neo-Marxists and dependency theorists invoked different forms of 'de-linking' from the capitalist world system, Lewis argued that to create a new international order it was necessary to increase the productivity of workers employed in food production in the developing world.[36] This would then alter the terms of trade, 'raise the prices of the traditional agricultural exports', and 'create an agricultural surplus that would support industrial production for the home market'.[37] In contrast with many of his contemporaries, who saw in indigenous agricultural smallholders an obstacle to modernisation and industrialisation, Lewis's strategy of economic development invested them – if supported by educational policies that increased their productivity – with the power to unleash structural transformations and alter the course of history.[38]

In 1955 Lewis published his most ambitious work: *The Theory of Economic Growth*. Despite the narrow claim that 'the central problem in the theory of economic growth is to understand the process by which a community is converted from being a 5 per cent to a 12 per cent saver', *The Theory of Economic Growth* is a book of dazzling complexity.[39] While identifying three main causes for the increase of output per head (the will to economise, increase in knowledge and capital accumulation), Lewis was keen to reiterate the mere analytical purpose of this distinction, and to show that in practice the causes of growth were overlapping and mutually constitutive.[40] At a time dominated by clashes between seemingly mutually exclusive approaches (e.g., balanced growth vs. unbalanced growth) and monocausal growth models, Lewis not only masterfully combined theory and empirical evidence to broaden the debate on the causes of growth, but also gestured at the ultimately non-economic nature of the 'causes of these causes'.[41] Even without amounting to a unified growth theory that could be easily summarised or tested econometrically, Lewis's book raised profound questions about the applicability of existing growth models to developing economies, while simultaneously striving for a more universal set of variables that could explain – or at the very least formulate hypotheses about – the historical trajectories of economies at different levels of development. Informed by Lewis's vast reading on economic history, sociology and anthropology, *The Theory of Economic Growth* is a *tour de force* embodying his belief that the economics of poor countries required a capacious and interdisciplinary approach. From this point of view, the volume did justice to its author's claim that, for breadth of scope and vision, it was the first comprehensive treatise on growth since 'John Stuart Mill's *Principles of Political Economy*, published in 1848'.[42]

The 1954 article and the 1955 book contributed to catapult Lewis to fame, making him one of the most visible experts on developing economies, while simultaneously helping carve out a new academic field that, while conversant and overlapping with growth economics, could depart from it in significant ways. During his time in Manchester, Lewis also produced some important policy reports for the Colonial Office and the United Nations.[43] These encapsulate Lewis's firm commitment to making his academic work relevant for the developing world, and testify to his distinctive capacity to weave theory and historical narratives into pragmatic policy prescriptions.

While achieving global visibility, Lewis's ideas were nurtured in the Victorian building in Dover Street that hosted the Faculty of Economic and Social Studies (figure 10.3). Lewis found a vibrant community of progressive scholars that included refugee economists such as Kurt Martin and Gisela Eisner (Lewis's only doctoral student at Manchester), and the anthropologist Max Gluckman (see Chapter 12 in this volume).[44]

Yet there is another, deeper sense in which Manchester shaped Lewis's understanding of development. The analytical narrative of labour moving from the 'subsistence' to the 'modern' sector mirrors a stylised account of the British Industrial Revolution, which found in Manchester one of its epicentres. Lewis's embrace of this

10.3 Dover Street c. 1960, with the Faculty of Economic and Social Studies on the left (previously the Manchester High School for Girls). Lewis worked there for the duration of his time at Manchester. Courtesy of the University of Manchester.

interpretation (challenged in contemporary economic historiography) was influenced by J. L. Hammond and Barbara Hammond's trilogy on British labour history between 1760 and 1832.[45] While accounting for the labour shift across sectors as key to capital accumulation, the Hammonds' works also taught Lewis that 'the industrial revolution had not raised urban wages'.[46] In the Hammonds' books, Manchester, 'not thrown up by the Industrial Revolution, but overwhelmed by it', plays a key role both as catalyst of change in the British 'dual economy' and as a reminder of the misery and poverty resulting from unregulated industrialisation and urbanisation.[47] While for Lewis 'the ugliness of towns or the impoverishment of the working classes' were not the unavoidable and 'necessary' costs of growth, Manchester's history simultaneously offered inspiration to develop new ideas on the dynamics of economic development, and a fertile – if difficult – ground to test them.[48]

Building communities in Moss Side and Hulme

In the 1950s Manchester still had several key features of a 'dual economy'.[49] These became more visible when assessed along racial and spatial lines. A significant part of Manchester's African and African Caribbean population lived in the Moss Side area, where they were deprived of educational opportunities and housing facilities, trapped in low-paying jobs or in unemployment, and victims of different forms of discrimination.[50] In 1948 the *Manchester Evening News* reported a spokesman from the National Coal Board saying that they would not 'employ coloured men in the mines where it can be avoided'; an enraged Lewis's response was to call for new legislation that could rectify racial discrimination in British labour markets.[51]

Lewis quickly became a pillar of the Mancunian African Caribbean community: his private correspondence contains requests for help (such as that for a desperate girl who had been abandoned by the child's father during her pregnancy), and evidence of his involvement in a wide range of activities, including the organisation of leafleting campaigns and musical events.[52] The situation of the African Caribbean communities was certainly dire: as it was put in a request for help from Christ Church, the Anglican parish church in Moss Side, the conditions of the thousands of people from Africa and the West Indies represented 'one of the most urgent social problems of our time'.[53] As a consequence, Lewis was committed to gather evidence on the African Caribbean experience of Manchester. The task fell upon Eyo Bassey Ndem, a Sierra Leonean anthropologist introduced to Lewis by Max Gluckman.[54] The resulting report portrayed a city in which significant tensions existed, for example in transport, as the Manchester public was not 'entirely in favour' of Black bus conductors.[55] An 'intense sex jealousy' also played a role: as many white people resented 'the idea of Coloured male workers fraternising with white girls', this was in turn used as an excuse to avoid employing people of African descent.[56] Discrimination was not limited to work places, as it pervaded recreational spaces such as hotels, restaurants and dance halls, not only in the city centre but also in Moss Side.[57] Finally, the prolonged experience of police brutality led the African Caribbean communities to 'distrust the police', and resulted in the exacerbation of 'feelings of frustration and exile'.[58]

Lewis identified in limited recreation, unemployment and lack of housing the key areas of intervention to improve the living conditions of the Moss Side working masses (the student population was in less need of support as their needs were partly catered for by the University, student associations and the British Council, among others).[59] Key to the resolution of these problems was, in Lewis's view, the support of already existing initiatives (such as a recreation centre set up by Christ Church), the collection of information, and the establishment and consolidation of relations with both sympathetic employers willing to give jobs to African Caribbean workers, and landlords who, outside the Moss Side area, 'would be prepared to give housing' to Black lodgers.[60] Legislative change did not exhaust the range of instruments required to rectify the racial injustice that was keeping the African Caribbean communities in abject poverty: instead, what was needed was a grounded, bottom-up attempt to intervene in the city and the communities inhabiting it. After finally convincing Manchester City Council, this vision underpinned the creation of two centres: Community House (in Moss Lane East) and the South Hulme Evening Centre (in Bangor Street) (fig 10.4).

While drawing inspiration from similar existing experiments in Birmingham, Lewis's scheme was more ambitious.[61] Not only did the centres offer education in a wide range of subjects, they were also geared towards providing 'vertical social capital' in the form of networks and connections with employers and social services, through which the Moss Side workers could overcome discrimination and access jobs and salaries more aligned with their effective skills.[62] Beyond formal education and networking opportunities, the centres also offered legal advice to make the workers more conscious of their rights and of the tools at their disposal to overcome discrimination. The South Hulme Centre proved to be short-lived, shutting down in 1955, but Community House kept Lewis's dream alive.

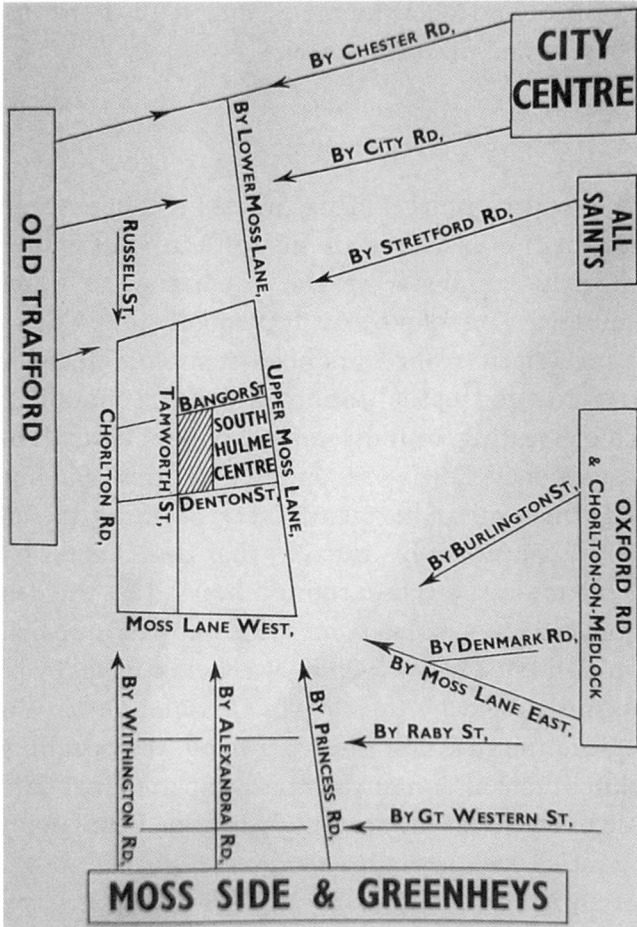

10.4 Map showing the location of the South Hulme centre, which Lewis was instrumental in founding. W. Arthur Lewis papers, Box 29, Folder 2.

Both had a positive impact on the economic opportunities of Moss Side communities.[63]

By the time Lewis left for the Caribbean in 1958, Manchester had exhausted its role in shaping his thought. But the impact had been a rich one: the city had provided him with a vibrant anti-colonial hub in which to rethink the entanglements of political and economic forms, with an analytical narrative to clarify the dynamics of economic development, and with a site to test his ideas on poverty and discrimination.

Conclusion

Manchester's 'unity of place' brings a certain coherence to Lewis's life and thought. It is easy to see in his work in Moss Side a concrete application of his emphasis on the importance of education for lifting communities out of poverty. It is equally tempting to identify in Lewis's prescriptions for Caribbean economic development an extension of his conceptualisation of dual economies, and the importance of creating an indigenous business class. Despite this apparent consistency, Lewis's legacies are multiple and fragmented. This is partly the result of his own choices, as he at times attempted to separate the aspects of his thought that this chapter has sought to unify. This tendency was certainly dictated by the desire to be taken seriously as a scholar who could provide impartial policy advice. The high esteem in which he was held equally by politicians such as Nkrumah and by the British Colonial Office shows that this was indeed the case. On the other hand, the bitter disappointment marking the end of many of these experiences confirms that he had little patience for the seemingly irrational and unpredictable nature of politics.

For contemporary economists, Lewis's work still provides a repository of intuitions and hypotheses in need of testing.[64] But specific aspects of his work have not aged particularly well. Lewis's simplistic take on environmental issues – and the idea that economic growth grants humankind more freedom by extending its mastery over nature – is a reminder of the Promethean nature of development economics in the 1950s and 1960s, when the imperatives of capital accumulation turned the environment into a blind spot.[65] On the other hand, Lewis's work offers useful counterpoints to the randomised control trials which have become pervasive in development economics.[66] Specifically, both his academic work and his politics can be read as an invitation to rescue a 'political economy' (not just 'economics') that is guided by common sense and historical awareness (albeit often very stylised) and is careful in its handling of empirical material, but that does not shy away from seeking to grasp 'the big picture'.

For this author, the entanglements of Lewis's life and thought with the city and the university of Manchester primarily raise questions on the nature of 'space' and 'place' in intellectual histories.[67] Manchester appears simultaneously as an epicentre of the imperial core and a gateway to anti-colonial worldmaking, as the institutional setting in which Lewis could flourish, and as embodiment and perpetuation of past economic and social inequalities. Is Manchester just a narrative device to impose a fictional coherence, or the material anchor and catalyst of Lewis's ideas? If asked, Lewis probably would have chosen not to answer. Not only did he know all too well that 'readers who need solutions tend to read them into a text, and to attribute to the writer positions which he does not hold', but he was also reluctant to become the subject of historical inquiry.[68] An enthusiastic scholar interested in writing about the development of Lewis's ideas and asking for both his approval and specific information received the following response: 'I have your letter [...] in which you invite me to collaborate to write my intellectual history, I am not willing to do so.'[69]

* * *

This chapter draws extensively on two excellent biographies: Robert L. Tignor, *W. Arthur Lewis and the Birth of Development Economics* (Princeton, NJ: Princeton University Press, 2006) and Barbara Ingham and Paul Mosley, *Sir Arthur Lewis: A Biography* (Basingstoke: Palgrave Macmillan, 2013). The latter carves a more prominent place for Manchester. I would like to thank Jesús Cháirez-Garza, Jacob Norris and Kerry Pimblott for their careful reading of this chapter. All errors remain my responsibility.

Notes

1 Norman Girvan, 'Sir Arthur Lewis – a man of his time, and ahead of his time – a public lecture', *Journal of Eastern Caribbean Studies*, 34:4 (2009), 82–3.

2 W. Arthur Lewis, 'Development economics in the 1950s', in Gerald M. Meier and Dudley Seers (eds), *Pioneers in Development* (Oxford: Oxford University Press for the World Bank, 1984), p. 132.

3 Manchester City Library (hereafter MCL), Papers of Ernest Simon, 1st Baron Simon of Wythenshawe, M11/13 Lewis, 'Note by Lord Simon – Prof Arthur Lewis, 24 July 1956'. Thanks to Stuart Jones for making me aware of this source and sharing it with me.

4 Thanks to Josephine Nevill for information on the lack of press coverage.

5 MCL, Ahmed Iqbal Ullah Race Relations Resource Centre, GB3228.34/1/8, 'Pan-African Congress, Press Release no. 7', p. 1.

6 The most exhaustive work on the Congress is Hakim Adi and Marika Sherwood, *The 1945 Manchester Pan-African Congress Revisited, with Colonial and ... Coloured Unity* (the report of the 5th Pan-African Congress), edited by George Padmore (London: New Beacon Books, 1995).

7 Adom Getachew, *Worldmaking after Empire: The Rise and Fall of Self-Determination* (Princeton, NJ: Princeton University Press, 2019), esp. ch. 3.

8 The Congress's resolutions are reproduced in Adi and Sherwood, *Manchester Pan-African Congress*, pp. 102–14.

9 W. Arthur Lewis, 'W. Arthur Lewis', in W. Breit and R. W. Spencer (eds), *Lives of the Laureates: Seven Nobel Economists* (Cambridge, MA: MIT Press, 1986), p. 12.

10 On Lewis's life and work until 1945, see Robert L. Tignor, *W. Arthur Lewis and the Birth of Development Economics* (Princeton, NJ: Princeton University Press, 2006), pp. 1–78, and Barbara Ingham and Paul Mosley, *Sir Arthur Lewis: A Biography* (Basingstoke: Palgrave Macmillan, 2013), pp. 1–84.

11 Ingham and Mosley, *Sir Arthur Lewis*, pp. 44–5.

12 Yoichi Mine, 'The political element in the works of W. Arthur Lewis: the 1954 Lewis model and African development', *The Developing Economies*, 44:3 (2006), 341, doi: 10.1111/j.1746-1049.2006.00020.x.

13 Lewis, 'W. Arthur Lewis', p. 12.

14 W. Arthur Lewis, *Labour in the West Indies: The Birth of a Workers' Movement with an Afterword 'Germs of an Idea' by Susan Craig* (1939) (London: New Beacon Books, 1977), pp. 17–43.

15 Ibid., pp. 45–9.

16 Ibid., p. 52.

17 For a detailed account, see Ingham and Mosley, *Sir Arthur Lewis*, ch. 7; Tignor, *W. Arthur Lewis*, ch. 7.

18 His disillusionment is recounted in W. A. Lewis, 'The agony of the eight', reprinted in *Journal of Eastern Caribbean Studies*, 23:1 (1998), 6–26.

19 W. A. Lewis, *Racial Conflict and Economic Development: The W.E.B. Du Bois Lectures 1982* (Cambridge, MA: Harvard University Press, 1985), p. 121.

20 Mine, 'Political element', especially pp. 346–51.

21 W. A. Lewis, *Politics in West Africa: The Whidden Lectures for 1965* (London: George Allen and Unwin, 1965), Acknowledgments (n.p.), pp. 49–55.

22 Theo Williams, 'George Padmore and the Soviet model of the British Commonwealth', *Modern Intellectual History*, 16:2 (2019), 531–59, doi: 10.1017/S1479244317000634.

23 Lewis, 'W. Arthur Lewis', p. 3; Federico D'Onofrio and Gerardo Serra, 'Arthur Lewis and the classical foundations of "development": economic history and institutional change', *Research in the History of Economic Thought and Methodology*, 37A (2019), 157–64, doi: 10.1108/S0743-41542019000037A012.

24 W. A. Lewis, 'Colonial development', paper presented at a meeting of the Manchester Statistical Society, 12 January 1949, p. 18, cited in Ingham and Mosley, *Sir Arthur Lewis*, p. 262.

25 W. Arthur Lewis papers (hereafter WAL), box 35, folder 2, 'The Victoria University of Manchester, Conditions of Appointment' [no date].

26 Frederick Cooper, 'Modernizing bureaucrats, backward Africans, and the development concept', in Frederick Cooper and Randall M. Packard (eds), *International Development and the Social Sciences: Essays in the History and Politics of Knowledge* (Berkeley: University of California Press, 1997), pp. 64–92.

27 Albert O. Hirschman, 'The rise and fall of development economics', in *Essays in Trespassing* (Cambridge: Cambridge University Press, 1981), pp. 1–24.

28 On the historiographical implications of this mixture, see Michele Alacevich and Mauro Boianovsky, 'Writing the history of development economics', in Michele Alacevich and Mauro Boianovsky (eds), *The Political Economy of Development Economics: A Historical Perspective*, supplement to *History of Political Economy*, 50 S1 (2018), 1–14, doi: 10.1215/00182702-7033812.

29 W. Arthur Lewis, *The Theory of Economic Growth* (London: George Allen and Unwin, 1955), p. 18.

30 Mauro Boianovsky, 'Arthur Lewis and the classical foundations of development economics', *Research in the History of Economic Thought and Methodology*, 37A (2019), 103–43, doi: 10.1108/S0743-41542019000037A009.

31 W. Arthur Lewis, 'Economic development with unlimited supplies of labour', *The Manchester School*, 22:2 (1954), 139–91, doi: 10.1111/j.1467-9957.1954.tb00021.x.

32 Ibid., pp. 140–1.

33 Ibid., p. 146.

34 Ibid., p. 171.

35 W. Arthur Lewis, *The Evolution of the International Economic Order* (Princeton, NJ: Princeton University Press, 1978).

36 Anthony Brewer, *Marxist Theories of Imperialism: A Critical Survey* (London: Routledge and Kegan Paul, 1980); Lewis, *Evolution of the International Economic Order*, p. 37.

37 Lewis, *Evolution of the International Economic Order*, p. 37.

38 Ingham and Mosley, *Sir Arthur Lewis*, p. 118.

39 Lewis, *Theory of Economic Growth*, pp. 225–6.

40 Ibid., p. 20.

41 Ibid., p. 11; D'Onofrio and Serra, 'Arthur Lewis', p. 158.

42 Lewis, *Theory of Economic Growth*, p. 5.

43 United Nations, *Measures for Economic Development of Under-Developed Countries: Report by a Group of Experts Appointed by the Secretary-General of the United Nations* (New York: United Nations, Department of Economic Affairs, 1951); *Gold Coast Government, Report on Industrialisation and the Gold Coast by Prof. W.A. Lewis* (Accra: Government Printer, 1953).

44 Ingham and Mosley, *Sir Arthur Lewis*, pp. 90–9.

45 J. L. Hammond and Barbara Hammond, *The Town Labourer 1760–1832: The New Civilisation* (London: Longmans, Green, 1919); *The Village Labourer 1760–1832: A Study in the Government of England before the Reform Bill* (London: Longmans, Green, 1919); *The Skilled Labourer* (London: Longmans, Green, 1919); Tignor, *W. Arthur Lewis*, pp. 90–3. For an influential contrasting interpretation of the causes of the Industrial Revolution, see Robert C. Allen, *The British Industrial Revolution in Global Perspective* (Cambridge: Cambridge University Press, 2009).

46 Lewis, 'W. Arthur Lewis', p. 14.

47 Hammond and Hammond, *Town Labourer*, pp. 42, 45.

48 Lewis, *Theory of Economic Growth*, p. 429.

49 Paul Mosley and Barbara Ingham, 'Applying the Lewis model in industrialised countries: W. Arthur Lewis and the dual economy of Manchester in the 1950s', *The Manchester School*, 84:1 (2016), 95–124, doi: 10.1111/manc.12087.

50 Ibid.

51 Mine, 'Political element', p. 346.

52 WAL, box 29, folder 5, 'Letter of Larry Walsh to W. Arthur Lewis, 8 October 1953'; 'Letter of W. Arthur Lewis, September 1953'; 'Letter of Lillian Samuels to W. Arthur Lewis' [undated].

53 WAL, box 29, folder 5, 'Letter of Revd. Canon T.F. Ethell (et al.) to W. Arthur Lewis' [undated].

54 Ingham and Mosley, *Sir Arthur Lewis*, p. 121.

55 WAL, box 29, folder 5, 'Memorandum by E.B. Ndem', p. 2.

56 Ibid.

57 Ibid., p. 4.

58 Ibid., p. 5. For a contemporary reflection on the impact of police racism on Manchester's Black communities, see K. Pimblott and K. Foale (eds), *A Growing Threat to Life: Taser Usage by Greater Manchester Police* (Manchester: Resistance Lab, 2020), https://resistancelab.network/our-work/taser-report/index.html (accessed 20 January 2024).

59 WAL, box 29, folder 5, 'Manchester Council for African Affairs, Agenda for Meeting on August 20th at 7.30 p.m., Methodist Hall, Oxford Road, Memorandum by Prof. Lewis', p. 1.

60 Ibid., pp. 2–3.

61 WAL, box 4, folder 1C, 'Letter of W. Arthur Lewis to the Chief Education Officer, Deansgate, Manchester, 28 October 1952'.

62 Mosley and Ingham, 'Applying the Lewis model', 110.

63 Ibid.

64 For example, Angus Deaton and Guy Laroque, 'A model of commodity prices after Sir Arthur Lewis', *Journal of Development Economics*, 71:2 (2003), 289–310, doi: 10.1016/S0304-3878(03)00030-0.

65 Lewis, *Theory of Economic Growth*, p. 421. For a critical analysis, see Matthias Schmelzer, Aaron Vansintjan and Andrea Vetter, *The Future is Degrowth: A Guide to a World Beyond Capitalism* (London: Verso, 2022). Thanks to Georg Christ for many enjoyable discussions on these issues.

66 For an introduction, see Abhijit V. Banerjee and Esther Duflo, *Poor Economics: A Radical Rethinking of the Way to Fight Global Poverty* (New York: Public Affairs, 2011). Pertinent critiques include Angus Deaton, 'Instruments, randomization, and learning about development', *Journal of Economic Literature*, 48 (2010), 424–55, doi:10.1257/jel.48.2.424; Ingrid H. Kvangraven, 'Nobel rebels in disguise: assessing the rise and rule of the randomistas', *Review of Political Economy*, 32:3 (2020), 305–41, doi: 10.1080/09538259.2020.1810886.

67 For example, Daniel S. Allemann, Anton Jäger and Valentina Mann, 'Introduction: approaching space in intellectual history', *Global Intellectual History*, 3:2 (2018), 127–36, doi: 10.1080/23801883.2018.1450614. Thanks to Mélanie Lindbjerg Machado-Guichon for making me aware of this.

68 Lewis, *Evolution of the International Economic Order*, p. 76.

69 WAL, box 4, folder 5D, 'Letter of William Darity Jr. to W. Arthur Lewis, 22 October 1984'; 'Letter of W. Arthur Lewis to William Darity Jr., 1 November 1984'.

11

Tools versus minds: two Manchester computing traditions

James Sumner

In the summer of 1998 the University of Manchester and Manchester City Council staged a grand celebration of the city's contributions to digital culture, featuring a commemorative concert by the Hallé Orchestra, festivals of electronic art and music, a dozen academic conferences, and public demonstrations of new prospects for managing work and social activity online. The festivities marked fifty years since the moment when, in the words of the promoters, 'what we now call the computer was born': on 21 June 1948 an experimental prototype nicknamed the 'Manchester Baby' became the first machine in the world to run a program stored in electronic memory. On 21 June 1998 a working replica of the long-dismantled Baby, painstakingly engineered using authentic parts, was inaugurated as the centrepiece of the Museum of Science and Industry's new Futures Gallery.[1]

A figure seldom mentioned during the celebrations was Alan Turing (1912–54), the iconic computer theorist and Second World War codebreaker who spent the last six years of his life at the University. Though Turing had not been involved in the Baby's creation, arriving only in October 1948, his work soon afterwards – in particular, his famous predictions about machine intelligence – seems an obvious feature of the Manchester computing story today. The reasons for Turing's exclusion lie partly in relatively straightforward departmental and interpersonal politics; yet there is a richer underlying story about conflicting visions of what counted as significant, and what counted as specifically Mancunian. The engineers who set the tone for the 1998 celebrations saw computers as fast, convenient tools for calculating or performing other well-defined tasks under human control. Turing, by contrast, saw the new technology as a gateway to exploring the very concepts of thought and understanding. Only in the twenty-first century has the University begun to define a history that accommodates both visions.

Inventing a computer

The 'first computer' is one of history's great red herrings. Innovators have been building automatic calculating machines of broadly increasing power and convenience since ancient times, employing whatever scientific and technical advantages came their way. During the Second World War, multiple research teams in Britain and the USA began to apply recent discoveries in electronics that promised huge increases in the speed of calculation, and perhaps the flexibility to perform very different tasks without laborious rewiring. Only in hindsight, after the first practical successes, did the emerging research community interpret these projects as steps on the road to a single outcome. One particular design approach, outlined by the US-based mathematician John von Neumann in 1945, became the dominant conceptual reference point, yet there were many ways to turn the concepts into practice.[2]

The 'invention of the computer', then, occurred over several years, in many places at once. In the United Kingdom, three sites in particular stand out: the government-funded National Physical Laboratory (NPL) near London, which recruited Alan Turing in October 1945 to work on a design which became known as the Automatic Computing Engine (ACE); the University of Cambridge; and Manchester, where the Victoria University's distinctive identity, marrying the broad subject coverage of traditional humanistic education to close engagement with the city's industrial culture, proved well suited to a new field that spanned mathematical logic at its most abstruse and electrical engineering at its most practical.[3]

Manchester already had an automatic calculation expert in Douglas Hartree (1897–1958), Professor of Applied Mathematics from 1929 to 1946. Although Hartree's own projects focused on analogue mechanical equipment, which worked on an entirely different principle, he was instrumental in fostering the first wave of digital computers in Britain and developing connections with research groups in the USA.[4] A yet more influential figure behind the scenes was Hartree's friend and colleague Patrick Blackett (1897–1974), Professor of Physics and a well-connected advisor on national science and technology policy. A champion of centrally planned, goal-directed science, Blackett swiftly recognised the value of electronic computers, and, like Hartree, saw strategic value in building up a project in Manchester with a distinct design focus from the other groups.

Blackett therefore persuaded the mathematician Max Newman (1897–1984), who had directed the top-secret Colossus codebreaking machine project during the war, to move to Manchester as Professor of Pure Mathematics in September 1945. He then encouraged Newman to seek a large grant from the Royal Society to build and staff a computing laboratory, and used his influence to ensure that the grant was awarded. Finally, at the end of 1946, Blackett managed the appointment as Professor of Electrotechnics (later renamed Electrical Engineering) of F. C. Williams (1911–77), a younger colleague who had worked with him on systems to be used with Hartree's analogue machines before moving into wartime radar research.[5] Though Williams was not a computer specialist, he had already made a discovery that would shape the character of Manchester computing.

One of the key challenges in computer engineering was to find a reliable, affordable way to store numbers and instructions without losing the advantage of electronic speed. Williams and his research assistant, Tom Kilburn (1921–2001), had ingeniously adapted the cathode-ray tubes used for radar displays so that the pattern of 'blips' on the screen came from computer data, with dots and dashes representing the zeroes and ones of binary code. Where other engineers began with a computer design and searched for components to make it work, Williams and Kilburn were in something like the

11.1 Panoramic view of the Manchester Mark 1 computer, composited from photographs taken in December 1948. Courtesy of the University of Manchester.

reverse position: they had a promising component, but needed to build a working computer around it to demonstrate its usefulness. The machine that ran in June 1948 was about the simplest possible that would allow a 'believable' demonstration.[6]

Following this success, however, the obvious goal was to expand the computer into something that could be used for serious computation, as Blackett intended. It was its prototypical character that inspired the nickname 'Baby': though the machine was five metres long and weighed around a tonne, it was much smaller than the full-scale 'Manchester Mark 1', which had largely taken shape by the end of 1948 (figure 11.1) and was completed in two stages the following year. This machine, in turn, served as the testbed for a yet larger model, designed for commercial sale by the electrical firm Ferranti, which worked closely with the University.[7]

In this incremental process, various milestones might have been nominated as the 'birth' of the Manchester computing phenomenon, but a preference for the Baby's first operation was seeded early. In September 1948 Williams and Kilburn reported in the journal *Nature* that a 'small electronic digital computing machine has been operating successfully for some weeks' in their laboratory.[8] In October an internal University report by Newman reduced the tangle of developments to the narrative that was to become conventional in Manchester: von Neumann's 1945 proposals were set to define future developments globally; Manchester had achieved the first implementation of the proposals that 'actually worked'; and the breakthrough had come the previous summer.

Newman's report also highlighted a significant divergence:

> There are two distinct fields of enquiry connected with large automatic computing machines. These are first, the engineering problem of designing the machine [...] secondly, the mathematical and logical problems of finding the best use of such machines and investigating their effect on the development of mathematics itself.[9]

Newman's gloss on the division was positive: the Manchester project happily combined both fields. Yet there was little synergy in practice between the two. The Williams–Kilburn group focused strongly on the first; Newman and his fellow mathematicians on the second. The arrival of Alan Turing, who had been mentored by Newman early in his career, sharpened the divergence.

Can machines think?

When Alan Turing resigned from the NPL and moved to Manchester, he was one of the few people in the world to have designed a computer; Manchester, for its part, was one of the few spots in the world where a computer had been designed. Many people assume, not unreasonably, that Turing came to Manchester to design computers. He did not. His attention was shifting to a new topic: given that computers *would* be designed, on an ever-increasing scale, what possibilities did their existence create? The NPL's ACE remained unbuilt, bogged down in administrative delays; the Manchester machine, though small and not of Turing's own design, was a real, practical computer that he could use as a testbed for his ideas. Though his role was officially to develop programming specifications and advise other users, Turing spent more and more time on independent research. He was pleased to find that local specialists in philosophy and social theory, notably Dorothy Emmet and the chemist-turned-philosopher Michael Polanyi, offered a receptive audience for his wide-ranging ideas.[10]

Characteristically, Turing's best-known work from this period, a remarkable speculative essay on future machine intelligence, appeared not in any mathematics or engineering periodical but in *Mind*, the pre-eminent journal of analytical philosophy, alongside papers on logic, the principle of sovereignty and the philosophy of language.[11] It was by no means a typical *Mind* paper. The journal's editor, the Oxford metaphysician Gilbert Ryle, loathed technicalities and rejoiced in authors who could make their point in two pages, seldom accepting more than twenty.[12] Of the twenty-eight pages of Turing's article, a quarter were devoted to explaining the principles of electronic digital computing, in far greater detail than the core argument required.

There was nothing technical, however, about the paper's opening section, with its memorable first line: 'I propose to consider the question, "Can machines think?"' Turing's choice of the term 'consider', rather than 'answer', was significant. Most authors who had responded to the question, he felt, had missed its limitations as a tool for understanding. One such was his Manchester colleague Geoffrey Jefferson, a prominent brain surgeon, who answered in the negative. The thinking brain, said Jefferson, was nothing like a computer, and was unique in its self-consciousness: a machine might be programmed to simulate a creative act, such as writing poetry, but only human minds could consciously experience creativity.[13] Turing objected that Jefferson was arguing in a circle: by choosing an entirely human-centred definition of 'thinking', he merely affirmed the obvious fact that machines were not human.[14]

To avoid circularities, Turing grounded his definitions in a practical test. He imagined a person holding remote text-based conversations with two unseen test subjects: a fellow human, and a computer programmed to respond like a human. The first person would then be challenged to identify the computer. If they could not do so, said Turing, to a better extent than they could identify imitations among people under the same conditions (his example involved distinguishing a woman from a man trying to write like a woman), then – whatever the basis of the computer's programming, and whether philosophers and brain specialists approved or not – most people would surely accept the term 'thinking' for the computer's behaviour.[15]

In taking the 'thinking machine' seriously, Turing set himself apart from the vast majority of computer theorists and designers. In 1946 hyperbolic newspaper coverage of ENIAC, one of the first large electronic computing machines, had seeded a popular belief in the United States that computers not only might, but already could, think like people.[16] Most experts found this characterisation both inaccurate and harmful: beyond creating impossible expectations, it fostered fears of mass unemployment, and of scientists pursuing a dehumanising agenda. This was a particular concern for Douglas Hartree, who sought to head off the 'electronic brain' interpretation before it could take root among British audiences.[17]

Hartree frequently stressed an argument which, he noted, had been anticipated in the 1840s by Ada Lovelace, in her account of Charles Babbage's extraordinary unbuilt design for a fully program-mable, gearwheel-driven 'Analytical Engine'. Such a machine, said Lovelace, would have 'no pretensions whatever to originate anything', but could do only 'whatever we know how to order it to perform'.[18] In the same way, said Hartree, an electronic computer might make decisions automatically, but only as directed by the programmers: 'use of the machine is no substitute for the thought of organising the computations, only for the labour of carrying them out'.[19] This was very much the orthodoxy among computer engineers, shared by Williams, Kilburn and their collaborators at Ferranti, notably Vivian Bowden, who created the first promotional materials for commercial computers.[20] The only prominent dissenter, in Manchester at least, was Turing.

Turing had begun to undermine Hartree's stance while still at the NPL. In one press interview, he had spoken blithely of a future, 'possibly in 30 years, when it would be as easy to ask the machine a question as to ask a man'. Hartree, quoted in the same piece, rebuffed any downgrading of the mind's authority in the most reso-nant terms possible in 1946: 'The fashion which has sprung up in the last 20 years to decry human reason is a path which leads straight to Nazism.'[21] In his *Mind* article, Turing characterises such dismissals under the heading of a '"Heads in the Sand" Objection: "The consequences of machines thinking would be too dreadful. Let us hope and believe that they cannot do so"', when there was every reason to suppose they might. Hartree was also wrong, said Turing, to invoke Lovelace's objection. Since the Analytical Engine was never built, Lovelace had pardonably taken it for granted that it could 'never do anything really new', but the practically realised electronic machines often produced surprising outputs.[22]

What Turing's article did not discuss is *how* to make a computer think, except in very general terms. Rather than trying to replicate an educated adult mind from scratch, he proposed creating a 'child-machine' that could learn – that is, rewrite its decision-making systems automatically – by interacting with human teachers or the world around it.[23] He offered no clues, however, on approaches to coding or data representation. This is perhaps surprising given that Turing's unpublished work reveals that he had made a start on the problem: a 1948 report to the NPL describes in embryo what would later be termed a 'neural network', a system of interconnected circuits that process signals from each other, and from external stimuli, in ways that might allow the system to evolve into an organisational structure capable of giving relevant responses.[24] This approach, which at the broad level lies behind many of today's AI techniques, was developed independently by later researchers; Turing's report remained unpublished and unknown for many years.[25]

Turing's admirers have often noted his lack of interest in publishing his discoveries or securing credit for his innovations.[26] To say that he did not always present his ideas to best effect would be an understatement. Though the *Mind* paper's opening is a model of clarity, the remainder often feels wildly digressive, and it is sometimes hard to tell whether Turing is being entirely serious. He dwells, for instance, on the need to define the terms of the experiment so that engineers tasked with inventing a 'thinking machine' cannot simply breed with each other and present their child as an invention. Closing the loophole by banning women engineers, he reports soberly, would not be sufficient, because an all-male team might invent human cloning, which would be a dazzling scientific feat, but not the right one. Yet it was on this peculiar foundation that Turing built the requirement that the problem must be considered in terms of digital computers alone, a stipulation that was entirely serious. A 1950s reader accustomed to desk calculators might find the constraint excessive; Turing, who himself had given a key definition of the universality of computer systems in the 1930s, was increasingly sure it was not.[27]

For the remainder of his short life, Turing explored how various novel problems might be tackled by existing or future machines. In 1951 the Ferranti computer was delivered to the University and housed in a purpose-built new building (figure 11.2): Turing had an

11.2 The Ferranti Mark 1 computer's console, 1951. Standing at right is Alan Turing; seated are two Ferranti engineers, Brian Pollard and Keith Lonsdale. This is thought to be the only photo of Turing with a computer. Courtesy of the University of Manchester.

office upstairs, as did Tom Kilburn, who increasingly oversaw day-to-day management of the computer. The two men did not work well together.[28] Turing was perhaps liable to forget that Kilburn was not a technician in a subordinate role; Kilburn, for his part, saw Turing's speculations as a distraction from serious research. Often, Turing worked from his home in Wilmslow, simulating potential computational approaches by hand. In his last years he became greatly interested in modelling mathematically how patterns and structural forms develop in plant and animal life, a project that has drawn increasing interest in the twenty-first century.[29]

Two reputations grow

The engineers were meanwhile pressing ahead with an agenda much more in keeping with Hartree's template: designing and building number-crunching computers of ever-increasing power, flexibility and speed. By the time of Turing's death, at the age of just 41 in 1954, they had successfully implemented a much faster and more efficient Mark 2 design: like the Mark 1, it was commercialised by Ferranti, under the name 'Mercury'. Around this time, the leadership passed from Williams, whose interests lay more in other areas of electrical engineering, to Kilburn, who by contrast had firmly identified himself as a computer specialist. Kilburn remained in overall charge of computer research at the University for the next quarter-century.

Although Kilburn's group addressed a growing range of topics in computing theory and practice, its particular reputation was for designing whole computers and bringing them into production. Its third machine, developed concurrently with the Mark 2, was smaller and more experimental, exploring the new possibilities of transistor technology. Its fourth was on another scale entirely: 'Atlas', developed 1958–62, was a 'supercomputer', intended to reaffirm Britain's status as a major computing power by outperforming every other machine in the world. The research behind these projects included a number of key conceptual developments that proved highly influential internationally, notably the earliest high-level programming language and the innovation of virtual memory. The developers' view of the 'Manchester University computers' as forming a distinct lineage was reflected in the name chosen for the fifth machine, completed in 1974: 'MU5'.[30] All five designs were either commercialised directly by local manufacturers, or strongly influenced commercial models.

In 1975 Simon Lavington, a member of the MU5 design team, compiled a short book, *A History of Manchester Computers*. This did not set out to be a history of Manchester *computing*: Lavington's aim was to tell the story of the five Kilburn-era machines, with some brief context on the wider picture both locally and nationally. It was Lavington who pinned down the anniversary of the Baby's first successful run, which had not been recorded by the original team: Williams had placed it in July 1948, but Lavington determined, by reference to a notebook kept by another engineer, Geoff Tootill, that the date must have been 21 June.[31]

The practice of commemorating this date at ten-yearly intervals began in a small way, with a lunch and cake-cutting ceremony for the thirtieth anniversary in 1978. An article in the *Manchester Evening News* set the template for later legend making, stating that the Baby 'paved the way for general-purpose use – and changed the world'.[32] Lavington today considers it very unlikely that any of the principal developers encouraged this framing: they defined success in terms of a pattern of achievement rather than a single early event, and in any case showed no interest in public fame.[33] The 'birthplace of computing' claim, however, had an obvious strategic appeal to promoters of both the University's and the city's reputation.

The 1988 commemoration was more extensive, featuring a dinner, lectures and an exhibition covering both history and ongoing projects (representing 'the next forty years').[34] Kilburn had by this point retired, but leadership had passed to his long-term collaborator Dai Edwards (1928–2020), who had been present from the beginning of the Manchester computer lineage. The events coincided with the inauguration of a new University building, shared between the departments of Computer Science and Electrical Engineering. A suite of cleanrooms, dust-free controlled environments for the highly sensitive preparation of silicon chips and related technologies, was prominently sited on the ground floor, as if to showcase the distinctively hardware-oriented flavour of Manchester computing.[35]

The golden anniversary, in 1998, was the grandest of all. Planning for the summer-long celebration began over two years in advance, in consultation with an enthusiastic city council.[36] Promotion aimed at general audiences erased the testbed nature of the Baby, suggesting that Manchester had won a 'race' to invent the computer against the NPL, Cambridge and unspecified Americans.[37] The disconnection between the computer scientists themselves and the mood of civic boosterism was summed up in a memorable interview given alongside the Baby replica by the broadcaster Tony Wilson, who, in an inaccurate but impassioned harangue, insisted: 'The academics who are fussing around, saying this is the first stored-program computer, it's bollocks. This is the *first computer*.'[38]

By this point, the 'Manchester computer' lineage had come to a natural close, as industry and researchers internationally moved away from individual large machines towards assemblages of smaller units that could be networked together as required. An 'MU6' had been developed in the 1980s, but the project was more of an overall design approach, leading to both a single stand-alone computer for local use and a processor intended for networking; there was no MU7.[39] Nonetheless, and notwithstanding the creation of research groups in formal methods and machine learning, the overall ethos of the Department of Computer Science still emphasised hardware and systems engineering more than comparable groups at other British universities.[40] Tools, rather than minds, defined the Manchester outlook – or so it seemed.

Public awareness of Alan Turing in the 1990s remained limited, but was quietly growing. The tale of Turing's posthumous reputation is almost as extraordinary as his life story, encompassing the diverse efforts of, among others, Turing's mother Sara, who, without prior writing experience, produced a book-length biography; several computer scientists in the USA who adopted Turing's universality concept as emblematic of their aims; Donald Michie, an artificial intelligence expert who had known Turing in his codebreaking days, and Brian Randell, one of the first historians of computing, who between them broke down the official secrecy around Turing's wartime role; Andrew Hodges, mathematician and Gay Liberation activist, whose 1983 biography remains among the most remarkable studies of any scientific personality; Hugh Whitemore, whose 1986 play *Breaking the Code* created a signature role for Derek Jacobi; John Graham-Cumming, whose 2009 petition for a government pardon for Turing's 'gross indecency' offences struck a public chord; and Barry Cooper, central coordinator of the international events marking Turing's centenary in 2012.[41] The heavily fictionalised feature film *The Imitation Game* (2014), starring Benedict Cumberbatch, cemented Turing's status as one of the best-known scientific figures of all time.

This exceptional rise had conspicuously little to do with Manchester, and less to do with its University. Turing's work had, in fact, been represented during the 1988 celebrations, through a performance of *Breaking the Code* by the University Stage Society, but not within the exhibition itself.[42] In the much larger 1998 programme, he was the focus of only one presentation: Jack Copeland, one of the more forthright advocates for Turing's legacy, discussed the pre-Manchester neural network research at a satellite meeting organised on behalf of the British Society for the History of Science by Jon Agar, the historian of technology who was then responsible for an ongoing project to develop a national computing archive in Manchester.[43]

The first permanent public commemoration in the city came in 1994, when the city council designated a new link road 'Alan Turing Way', after a recommendation from its Gay Men's Subcommittee.[44] This was followed in 2001 by a bronze sculpture of Turing in the city centre (plate 5), following a campaign initiated by Richard Humphry, a barrister in Stockport, who learned of Turing's story through a local production of *Breaking the Code*.[45] Plans for the monument emphasised the appropriateness of its location in Sackville Park, between the city's Gay Village and 'the university science buildings'[46] – but the university in question was UMIST, until 2004 a separate institution from the Victoria University where Turing had worked.

Why did the Department of Computer Science show so little interest in Turing's emerging and enthralling story? Most recollections cite the influence of Tom Kilburn, who, though long since officially retired, remained an active presence, contributing with particular enthusiasm to the Baby replica project in the 1990s.[47] It was not, however, that Kilburn refused to sanction any mention of Turing: the two had not got on, but they were hardly bitter rivals. It was more a case of the department's self-image being very firmly bound up with the 'Manchester computer' lineage that began with the Baby. The fiftieth anniversary commemorations were coordinated by the husband-and-wife team of Brian Napper and Hilary Kahn, both professors in the department, and both former contributors to the Kilburn machine projects.[48] The pair were also chiefly responsible for the impressive heritage displays and artworks that decorated the University's main Computer Building, which became the Kilburn Building following Kilburn's death in 2001: again, these chiefly celebrated the Williams–Kilburn team, the Baby, and its successor machines. To mention Turing's name in this context was not anathema, but it was locally accepted to be irrelevant: Turing was interesting, but he was primarily a mathematician, or a philosopher of mind, or otherwise outside the tradition of Manchester computing.

The two traditions reconciled?

Unsurprisingly, the relative marginalisation of Turing at the University did not survive the growing upsurge of global interest. 2007 saw the first permanent commemoration on campus, when a major new teaching and research facility was named the Alan Turing Building, although it primarily houses mathematicians and physicists. The School of Computer Science paid a significant tribute in 2012, however, with the Alan Turing Centenary Conference, an international event organised by one of its professors, Andrei Voronkov, which brought together some of the most high-profile researchers in the world.[49] More recently the Kilburn Building itself has seen a significant overhaul of its heritage presentation, overseen by Jim Miles, a former head of school: Turing is well represented, as, incidentally, are Hartree and a more diverse range of researchers from later periods. A social space for students created in 2019 is named the Turing Lounge.

Some current research, meanwhile, interestingly combines elements of both traditions. Steve Furber, who joined the University in 1990, is primarily a hardware engineer, with an international reputation for designing stand-alone microcomputers and the ARM series of processors that power most smartphones and lightweight computers today, and his projects at Manchester have chiefly concerned processor design. At the same time, since the early 2000s, his group has been exploring the possibilities of using extremely large numbers of ARM processors to create a neural network, similar in overall principle to the system Turing described in 1948.[50] Turing had once spoken of his desire to 'build a brain'; Furber reported his project as having reached 'the scale of a mouse brain' in 2019.[51]

Neural networks are also the basis of the 'large language model' (LLM) systems that have reshaped public awareness of artificial intelligence since around 2022, with the launch of seemingly sentient question-and-answer systems such as OpenAI's ChatGPT. LLMs in fact bear only a very limited resemblance to the 'child-machines' Turing imagined learning by interaction with the world: for the most part, they simply identify patterns across huge volumes of human-generated text and imitate them. Turing would perhaps have accepted that Hartree's objections applied in their case, since the claim that they do not understand what they are saying is fairly easy to document in terms of their observable behaviour, given their notorious tendency to offer answers that are plausible at first glance, but clearly wrong when checked.

What does the future hold? A common myth, based on a misreading of a line in Turing's 1950 paper, is that he believed machines would reliably imitate people by the year 2000: in fact, Turing stated in 1952 that it would be 'at least 100 years' before the idea became a reality.[52] Perhaps the Baby's centenary in 2048, or the University's next major anniversary in 2074, will be witnessed by a human-scale learning machine; or perhaps, as seems more likely, the research capacities of Manchester and the world will have more pressing priorities to deal with – a scenario Turing, with his restlessly interdisciplinary spirit, would have taken in his stride. Time will tell.

Notes

1 Brochures from the celebrations are preserved in the University Archives, University of Manchester Library (hereafter UML), MUC/9/39. Quotation is from 'Computer 50: the University of Manchester celebrates the birth of the stored-program computer', p. 3. See also the anniversary meeting website: curation.cs.manchester.ac.uk/computer50/www.computer50.org/mark1/celebrations.html (accessed 28 March 2024); Jon Agar, 'Digital patina: texts, spirit and the first computer', *History and Technology*, 15:1–2 (1998), 121–35; and Jon Agar, Sarah Green and Penny Harvey, 'Cotton to computers: from industrial to information revolutions', in Steve Woolgar (ed.), *Virtual Society? Technology, Cyberbole, Reality* (Oxford: Oxford University Press, 2002), pp. 264–85.
2 John von Neumann, 'First draft of a report on the EDVAC', 1945, web.mit.edu/STS.035/www/PDFs/edvac.pdf (accessed 28 March 2024). See also Thomas Haigh, Mark Priestley and Crispin Rope, 'Reconsidering the stored-program concept', *IEEE Annals of History of Computing*, 36:1 (2014), 4–17.
3 For the University's ethos, see H. B. Charlton, *Portrait of a University, 1851–1951* (Manchester: Manchester University Press, 1951), pp. 53–109.
4 For Hartree's time in Manchester, see Charlotte Froese Fischer, *Douglas Rayner Hartree: His Life in Science and Computing* (Singapore: World Scientific, 2003), pp. 73–160.
5 Mary Croarken, 'The beginnings of the Manchester computer phenomenon: people and influences', *IEEE Annals of the History of Computing*, 15:3 (1993), 12–14; Simon Lavington, 'Early days of computing at Manchester: Max Newman's Royal Society project, 1946–1951', *IEEE Annals of History of Computing*, 44:2 (2022), 22–3; David Anderson, 'Patrick Blackett: physicist, radical, and chief architect of the Manchester computing phenomenon', *IEEE Annals of the History of Computing*, 29:3 (2007), 82–5; Froese Fischer, *Hartree*, pp. 145, 152.
6 Interview with Tom Kilburn by David Reid, 2 April 1998. My thanks to Simon Lavington and Jim Miles for this reference.
7 Simon Lavington, *A History of Manchester Computers* (Manchester: NCC, 1975), pp. 12–20.
8 F. C. Williams and Tom Kilburn, 'Electronic digital computers' (letter), *Nature*, 162 (1948), 487.
9 UML, NAHC/MUC/2/C/2, Max Newman, 'The University of Manchester: Arrangements for Royal Society Computer Machine Laboratory', 15 October 1948.
10 Jonathan Swinton, *Alan Turing's Manchester* (Manchester: Infang, 2019), pp. 87–94.
11 Alan Turing, 'Computing machinery and intelligence', *Mind*, 1950:4 (1950), 433–60.

12 D. W. Hamlyn, 'Gilbert Ryle and *Mind*', *Revue Internationale de Philosophie*, 223:1 (2003), 5–12.
13 Geoffrey Jefferson, 'The mind of mechanical man', *British Medical Journal*, 25 June 1949, 1110.
14 Turing, 'Computing machinery', 445–7.
15 Ibid., 433–5.
16 C. Dianne Martin, 'The myth of the awesome thinking machine', *Communications of the ACM*, 36:4 (1993), 120–33.
17 James Sumner, 'The United Kingdom: going it alone?', in Dick van Lente (ed.), *Prophets of Computing: Visions of Society Transformed by Computing* (New York: ACM, 2022), pp. 122–4.
18 Quoted in Douglas Hartree, *Calculating Instruments and Machines* (Urbana: University of Illinois Press, 1949), p. 70. Emphasis original.
19 Douglas Hartree, 'The ENIAC, an electronic computing machine', *Nature*, 158 (12 October 1946), 505; *Science Survey*, BBC Home Service, 11 December 1946, transcript in BBC Written Archives. My thanks to Allan Jones for the latter reference.
20 James Sumner, 'Defiance to compliance: visions of the computer in postwar Britain', *History and Technology*, 30:4 (2014), 309–33.
21 '"ACE" will speed jet flying', *Daily Telegraph*, 8 November 1946, 5. See also Bernardo Gonçalves, 'Lady Lovelace's objection: the Turing–Hartree disputes over the meaning of digital computers, 1946–1951', *IEEE Annals of History of Computing*, 46 (2024), 6–18.
22 Turing, 'Computing machinery', 444, 450–1.
23 Ibid., 454–9.
24 Reproduced in Jack Copeland (ed.), *The Essential Turing* (Oxford: Clarendon Press, 2004), pp. 410–32.
25 Copeland (ed.), *Essential Turing*, p. 409; Christoph Teuscher, *Turing's Connectionism: An Investigation of Neural Network Architectures* (London: Springer, 2002).
26 James Sumner, 'Turing today', *Notes and Records of the Royal Society*, 66 (2012), 300.
27 Turing, 'Computing machinery', 435–6.
28 The most detailed characterisation of the tension between Turing and Kilburn appears in the recollections of Geoff Tootill: interviews conducted by Tom Lean, 2009–10, National Life Stories C1379/02.
29 Jonathan Swinton, 'Watching the daisies grow: Turing and Fibonacci phyllotaxis', in Christof Teuscher (ed.), *Alan Turing: Life and Legacy of a Great Thinker* (Berlin: Springer, 2004), pp. 477–98; Swinton, *Alan Turing's Manchester*, pp. 127–41.
30 The standard overview of these developments remains Lavington, *History of Manchester Computers*. A second edition (Swindon: British Computer Society) was published in 1998.
31 Lavington, *History of Manchester Computers*, p. 10; cf. F. C. Williams, 'Early computers at Manchester University', *The Radio and Electronic Engineer*, 45 (1975), 329.

32 Brian Hope, 'Eureka! It was time to celebrate', *Manchester Evening News*, 21 June 1978, 13; photographs in UML Archive, MUC/8/65.

33 Simon Lavington, personal communication, 24 February 2024.

34 UML, MUC/8/69 and 70; MUC/9/35 and 36.

35 My particular thanks to Jim Miles for his advice on this point (personal communication, 22 February 2024).

36 UML, MUC/9/39, James Grigor, 'The Universal Machine', March 1996.

37 For instance, UML, MUC/9/39, 'Computer 50: the University of Manchester celebrates the birth of the stored-program computer' (brochure, 1998).

38 A transcript of the interview is at https://curation.cs.manchester.ac.uk/digital60/www.digital60.org/media/interview_tony_wilson/transcript.html (accessed 28 March 2024).

39 'MU6G instruction set manual, issue 2' (1982; re-set edition, 2019), archive.org/details/mu6ginstsetum1982/ (accessed 28 March 2024); J. V. Woods and A. J. T. Wheen, 'MU6P: an advanced microprocessor architecture', *Computer Journal*, 26 (1983), 208–17.

40 The shape of the department's research profile by the 1990s may be judged from archived versions of its original website: see, for instance, web.archive.org/web/19970330034751/http://www.cs.man.ac.uk:80/index.html, archived 30 March 1997.

41 Andrew Hodges, *Alan Turing: The Enigma of Intelligence* (London: Burnett, 1983), pp. 530–40 surveys Turing's reputation up to 1977. For US computer scientists' commemorative efforts, see Edgar Daylight, 'Towards a historical notion of "Turing—the Father of Computer Science"', *History and Philosophy of Logic*, 36:3 (2015), 205–28. For later commemorations, see Jonathan Bowen and Jack Copeland, 'Turing's legacy', in Jack Copeland et al. (eds), *The Turing Guide* (Oxford: Oxford University Press, 2017), pp. 461–74.

42 UML, MUC/9/35, letter from J. R. Gurd dated April 1988.

43 Jon Agar, personal communication, 8 February 2024. The meeting programme is preserved at web.archive.org/web/20000414072844/http://www.mailbase.ac.uk/lists/mersenne/1998-05/0021.html (accessed 28 March 2024). Copeland's talk was based on work published as B. Jack Copeland and Diane Proudfoot, 'On Alan Turing's anticipation of connectionism', *Synthese*, 108 (1996), 361–77.

44 Jonathan Swinton, 'If I walked that way I'd get arrested: the birth of the Alan Turing Way', 2 October 2019, www.manturing.net/manufacturing-blog/2019/10/2/if-i-walked-that-way-id-get-arrested-the-birth-of-the-alan-turing-way (accessed 28 March 2024); Jonathan Swinton, personal communication, 18 January 2024.

45 'Alan Turing: replicating the enigma', Manchester Archives+ blog, 26 August 2012, https://manchesterarchiveplus.wordpress.com/2012/08/26/alan-turing-replicating-the-enigma/ (accessed 28 March 2024).

46 'Alan Mathison Turing, 1912–1954' (flyer), 2001, in Turing Memorial Fund archive, Manchester Archives+, M821, https://www.flickr.com/photos/manchesterarchiveplus/7832054370/in/photostream/ (accessed 28 March 2024).

47 Maurice Wilkes and Hilary Kahn, 'Tom Kilburn CBE, FREng, 11 August 1921–17 January 2001', *Biographical Memoirs of Fellows of the Royal Society*, 49 (2003), 295.

48 See Janet Abbate's interview with Hilary Kahn conducted in 2001, https://ethw.org/Oral-History:Hilary_Kahn (accessed 28 March 2024).

49 The conference website is preserved at https://curation.cs.manchester.ac.uk/Turing100/www.turing100.manchester.ac.uk/ (accessed 28 March 2024).

50 Steve Furber and Petruţ Bogdan (eds), *SpiNNaker: a Spiking Neural Network Architecture* (Hanover, MA: Now, 2020).

51 Ibid., p. x. Donald Bayley, an engineer who worked with Turing around 1944, recalled the mention of brain-building in interviews: see, for instance, Hodges, *Alan Turing*, p. 290.

52 Diane Proudfoot, 'The Turing test—from every angle', in Copeland et al. (eds), *Turing Guide*, p. 292. Turing made this estimate in a BBC radio discussion: transcript at Turing Digital Archive, AMT/B/6, turingarchive.kings.cam.ac.uk/publications-lectures-and-talks-amtb/amt-b-6 (accessed 28 March 2024), p. 6.

12

Social anthropology at Manchester and the study of modernity

Katherine Ambler

In 1949 the University of Manchester established a Department of Social Anthropology. It was a dynamic period for the University's Faculty of Social Studies: Social Anthropology was to join growing departments of Economics and Government as part of a project to develop the social sciences. The head of the new department, Max Gluckman, was South African by birth and had previously worked as director of the Rhodes-Livingstone Institute (RLI) in British Central Africa. His remit at Manchester was to develop a programme of research into 'modern' societies, especially Britain itself. While the new department would retain links with the RLI, it swiftly established a local programme of research, addressing questions ranging from the social impact of deindustrialisation, to the dynamics of the factory floor, to the new realities of social mobility.

Shaping an intellectual agenda

Anthropology at Manchester in Gluckman's era is probably best known for the work done in connection with the RLI. In the 1950s Manchester scholars published ethnographic monographs on British Central Africa which shared common themes and approaches – a focus on 'modernising' forces such as industrialisation and urbanisation, an interest in the social function of conflict, and the use of case studies to illustrate broader arguments, a technique they called

12.1 Max Gluckman, probably c. 1968. From *Biographical Memoirs of Fellows of the British Academy*, 1976, courtesy of the British Academy.

'extended-case analysis'.[1] Recognising these commonalities, the idea of the 'Manchester School of Social Anthropology' was first constructed by fellow anthropologist Mary Douglas in a 1959 book review.[2] The image of the 'Manchester School' that appears in the literature is of a tight-knit group of Africanists, hostile to outsiders, dominated by the charismatic Gluckman, bound by sympathy for far-left politics.[3] This chapter, however, does not focus on this Africanist strand of work, which is already widely covered in existing literature. Instead, it highlights the department's work on modern life at home in Britain, as well as the large-scale research project into Israel which formed the centrepiece of the final years of Gluck-man's career and life.

In early 1949 the University of Manchester advertised a Readership in Social Anthropology. The job announcement required that 'can-didates should be interested both in Modern and in Primitive Societies, but their main interests may lie in either field'.[4] It was Raymond Firth, Professor of Social Anthropology at the LSE, who recommended Max Gluckman as a potential candidate.[5] At this point, Gluckman was a lecturer at the University of Oxford. Gluck-man sufficiently impressed the interview panel that, according to their notes, 'the opinion was strongly expressed that an endeavour should be made to acquire the services of Dr Gluckman and that ... a recommendation should be submitted for his appointment as Chair'.[6] Writing to his former student Clyde Mitchell, who was then employed as a research officer at the RLI, Gluckman said that 'Manchester is keen on developing the study of modern communities in England'. He added that he had informed the committee that such research was only being done in the US and that he himself 'had given my life to primitive communities and would not be able to do that kind of work'. Instead, he had suggested that an additional appointment be made to develop this line of research.[7]

A later comment, also to Mitchell, explaining how this additional lectureship would be used points to the intentions of the University in creating the Anthropology Department: 'It will be specifically for the study of modern communities (western family and kinship, neighbourhoods, factories, etc.) ... to provide the academic background and research, with its equivalent of colonial administrators in the people we call social administrators – welfare workers, social workers, personnel managers, etc.' The whole of the University, he noted, was 'keen' on the development of such work.[8] Gluckman's work in anthropology was intended to fit in with other work on social administration being done in the faculty by the likes of Ely Devons, Arthur Lewis and Bill Mackenzie. While the faculty had apparently accepted that Gluckman himself preferred to continue working primarily on 'primitive communities', it was integral to the new

12.2 The Roscoe Building at the University of Manchester in 1965, shortly after its completion. In front of it is the temporary building in which the Social Anthropology Department was housed in the late 1950s/early 1960s. Gluckman called it 'Stalag 17, beyond the Urals'. The department subsequently moved into the Roscoe Building. Courtesy of Manchester Libraries, Information and Archives, M64038.

department from the beginning that anthropology should tie in with other work at Manchester geared towards the local community and towards training people to work with and for government administration. Gluckman was resentful when its growth meant that his department had to leave the Dover Street building occupied by most of the faculty and occupy first a temporary building on Brunswick Street, and later the new Roscoe Building.

As well as contributing to the University's goals of serving the local community and producing graduates who could staff the growing welfare state, Gluckman also wanted the new department to establish itself as a credible alternative to the older, more prestigious anthropology departments in Oxford, Cambridge and London. Manchester's anthropologists would indeed encounter a certain amount of snobbery from their peers. Bill Epstein, who worked in the department in the 1950s, claimed, for example, that members of the Manchester department were referred to as 'the cloth cap boys' when they attended Association of Social Anthropologists (ASA) meetings in London.[9] Reflecting on his time serving as chairman of the ASA, Gluckman would later argue that he had felt that he 'had to stay on in order to represent the Universities outside Oxford and Cambridge and London against Southern domination'.[10] Although he himself had been at Oxford as both student and lecturer, Gluckman was quick to adopt an adversarial approach towards his alma mater and champion what he saw as the more innovative work being undertaken by Manchester's anthropologists.

Gluckman wanted the new department to make a distinctive intellectual contribution to the field of social anthropology. In 1950 he claimed that Manchester would soon be 'the best school in England, it will make social anthropology in the future'. He dismissed its rivals and announced confidently that 'Manchester will become ... preferred to London and Oxford and Cambridge by a number of people, fairly soon ... We shall be the only intelligent self-conscious school in Britain which can do anything.'[11] By 1957 Gluckman was writing in a letter to one of his protégés, Ronnie Frankenberg, that he was considering writing a series of popular books on anthropology on the basis that 'we need books of this kind within British anthropology especially out of the Manchester School'.[12]

This Manchester style of anthropology would be grounded in contemporary problems, well placed to assess life in Britain itself, and could offer insights to scholars across a range of disciplines. It also embraced an inclusive vision of the subject that incorporated sociology and industrial research alongside the more standard anthropological studies of villages in Africa or Asia. Indeed, the department's full name was the Department of Social Anthropology and Sociology. Initially there was little differentiation between the two; staff were primarily anthropologists by training who were interested in sociological methods. In the 1950s it was sociology that was often seen as somewhat suspect and unproven, while anthropology was still widely considered the more credible approach to social analysis: this was why Manchester had recruited Gluckman, an anthropologist, rather than a sociologist to establish the new department.

Manchester's anthropologists were interested in comparison: the idea that comparing an aspect of social life in different places could generate new insights. This would prove significant for the studies of Britain that were undertaken in the 1950s: was life in a British town really all that different to life in, say, southern Africa? Or was it the case that custom, ritual and kinship were important everywhere? By shifting the focus from classic non-Western anthropological sites to metropolitan Britain, Manchester's social anthropologists were pushing against established definitions of who anthropologists were and what they studied. In this, the department reflected a period of fluidity in the social sciences in Britain, when it was possible to envisage a more encompassing model of social anthropology. They also sought to demonstrate the practical value of the discipline, which had the potential to contribute to conversations about contemporary challenges such as industrial unrest, social mobility, urban life and social cohesion.

Anthropologies of modern Britain

The first studies of British life would be more the product of circumstance than conviction, but did provide early demonstrations of the value to be found in shifting anthropology's focus homeward. Ronnie Frankenberg was denied access to first Northern Rhodesia, then Uganda, then the West Indies for his PhD fieldwork.[13] The reason for this denial, which was not officially given to Frankenberg, was that he had been identified as having been a member of the Communist Party in 1950.[14] Needing to find a more straightforward location for research, Frankenberg shifted the framework for his project to North Wales. In a draft proposal on how this revised scheme would work, Frankenberg wrote that he hoped to 'throw light on the problem of what constitutes a "community" in industrial society'.[15] The result of Frankenberg's PhD research would be published in 1957 as *Village on the Border*, exploring the tensions between Welsh and English residents of the village, and discussing how inhabitants were struggling to uphold their sense of tradition in the face of economic decline caused by the closure of local industry.[16]

This early effort to try out anthropology in Britain was also being attempted in a more central location: London itself. Derek Allcorn had joined the department with PhD funding from the Medical Research Council to study the social development of young men in Acton. His thesis highlighted the significance of male peer groups and the powerful role they played in setting norms for their members, overriding those of the family and other communities.[17] Allcorn's thesis would be notable for being one of the very few of the department's from the 1950s not to be published as a book.[18] Frankenberg later suggested that this was because publishers were then uninterested in such British ethnographic studies: those that were published tended either to be 'generalized' (such as the work of Richard Hoggart) or offered 'celtic or bucolic appeal' (such as his own study of Wales).[19] In fact, Frankenberg had himself struggled to get his book into print: it had been turned down by Manchester University Press, which had published much of the Anthropology Department's output.[20] It was apparently difficult for some to accept that a study of a London suburb or a Welsh village had the same appeal as the standard distant village location.

Following these ad hoc initiatives in the first half of the 1950s to develop British research projects, from the mid-1950s there would be a concerted effort to put together more ambitious schemes. Bill Watson developed a plan to explore the topical theme of social mobility. In an application to the Ford Foundation for financing, Gluckman noted that 'for some time I have been discussing with my colleagues in the Department of Economics and Government the possibility of our undertaking a joint study of urban life in North-West England'.[21] The plan was finally being put into motion thanks to Watson's vision: a large-scale study of social mobility in the Lancashire town of Leigh. The proposal was based on a team of research workers in Leigh focused on individual questions, but with 'co-ordinating analysis' also envisaged to pull the scheme together.[22] The Nuffield Foundation funded the project.[23] It would run until 1964, when Watson left to take up a post in the United States. However, despite encompassing a significant proportion of the department's time and personnel, it had little impact, since team members apparently struggled to find appropriate research subjects or develop methodologies that would suit their ambitious aims.[24] The Leigh Social Mobility Study remains something of a mysterious failure, albeit one that was successful in bringing research funding into the department for a number of years.

Factory studies

A much more successful anthropological research scheme that ran in the 1950s and early 1960s would be known as the 'Manchester Factory Studies'. These were inspired by the visit of George Homans, Harvard sociologist, to the department; Homans had worked with Elton Mayo and had written about the 'Hawthorne Experiments' that Mayo had overseen in a US factory in the 1920s and 1930s.[25] Homans's visit coincided with the announcement that the British Department of Scientific and Industrial Research (DSIR) was seeking proposals for social research into industrial problems. At Manchester, a plan was developed to restudy aspects of the Hawthorne research in a more explicitly anthropological framework. This involved a team of researchers not merely observing the behaviour of workers but becoming workers themselves. Researchers were employed by the factories they were studying and worked on the production line, with their own responsibilities and targets to meet. Workers, management and union officials knew that they were there to undertake research; it had been agreed that failure to disclose their academic status was too close to spying. In addition, while their research was grounded in detailed personal experience, it was contextualised by the kind of data – on wages, for example – that could only be obtained by working with the owners and officials of the factory.

One of the conclusions of the team's work was that broad generalisations about worker behaviour and factory life were not possible because the situation varied so widely from factory to factory. What they did instead was offer detailed examinations of both the physical and mental hardship of industrial labour and insight into the everyday conflict and struggle on the shop floor that marked workers' lives. Two of this team would ultimately publish books recounting their research.[26] Gluckman himself became increasingly involved with the DSIR, serving on its Human Sciences Committee from 1957 until the DSIR was closed in 1965. In a retrospective history of Manchester's work on Britain published in 1982, this research was described as probably 'the most enduring and specific of the legacies that Gluckman left to British sociology'.[27]

Alongside these pioneering research projects, the department's teaching also reflected the full range of interests of its staff members. Students enrolled on anthropology or social studies degrees at Manchester could also take a course on industrial sociology in their second year. Gluckman remarked that the module was 'a course good in itself but also valued by employers'.[28] In 1958 this industrial sociology module was being taught by three members of the Factory Studies team, who had two seminar groups each. Students who had completed this module could then take a follow-up in field sociology, which was based on in-depth discussions of industrial sociological monographs; this was taught by Tom Lupton, the Factory Studies lead. One of the department's other popular modules was 'British Social Groups': students could choose to specialise in either urban or rural communities. In contrast, the module 'Ritual Systems of Pagan Africa' only managed to attract a single student one year.[29] It appears to have been possible to graduate with a degree in anthropology from Manchester with only minimal reading about the classic sites of anthropological research, such as colonial Africa. In a 1960 letter to Manchester's vice-chancellor, Gluckman pointed out that 'in no other University are social anthropology and sociology fully integrated'. He was proud that, at Manchester, social anthropology and sociology were taught with equal status to economics and politics, creating 'a degree in social science which is unique in Britain'.[30]

Israel: community and modernity

In Manchester's first decade, Gluckman had succeeded in developing a strong body of research and teaching that drew on the approaches and ideas of both sociology and social anthropology, encouraging members of the department to range across the two disciplines, and extending the traditional remit of social anthropological research to encompass studies of Britain itself. The department had led the way in Britain in arguing both that African cities and industrial sites were 'modern' in the same way as their Western counterparts, and that Western societies could be studied in the same way as the usual anthropological village sites. Such claims had become increasingly politicised because, as the processes of decolonisation unfolded, decisions about where anthropologists could study were the subject of public debate and official intervention.[31] Gluckman had always been open to new avenues of funding that would enable him to expand his department and, in the 1960s, an opportunity presented itself to pioneer anthropological research in a new location where the authorities were happy to accept anthropologists and where the 'traditional' and the 'modern' could both be found: Israel.

The Bernstein Israeli Research Trust (BIRT) project funded ten researchers to undertake anthropological studies of communities in Israel between 1965 and 1970.[32] British media magnate Sidney Bernstein, founder of Granada Television, funded the project, following a visit to Israel that had inspired him to support research in the country. Bernstein also had an interest in supporting universities in the north of England: indeed, he informed Gluckman that 'he was not prepared to put any money into London, Oxford or Cambridge, since he thought they got enough money'.[33]

Israel offered the perfect opportunity to demonstrate the value of anthropological approaches to complex societies. The country was famous for its communal agricultural projects such as the kibbutzim – exemplars of the kind of small communities studied by anthropologists. At the same time, Israel was a self-consciously 'modern', highly complex new state engaged in a process of nation building. Achieving this required absorbing immigrants from all over the world, integrating a melting pot of languages, cultures and religious traditions – while also grappling with the status of Palestinians and fraught relationships with its Arab neighbours. Anthropology, Gluckman and his team believed, was well placed to support such nation-building efforts: ethnographic research could analyse the relationships between citizens and the state, the conflict between religious and secular culture, and the best ways of integrating immigrants and building a modern Jewish state.

In 1966, in a report produced by Manchester's Faculty of Social Studies, it was noted that the department had a 'deep and specialist interest in Israel, which, in terms of numbers of specialists and the research effort involved, is probably unequalled at any other University in Britain'.[34] The initial intention had been to study questions related to the integration and adaptation of immigrants in Israel. However, while recent immigrants would indeed be the subject of some of the projects, others were studying more established communities: moshavim (cooperative smallholdings), kibbutzim (communal settlements), so-called 'development towns', and communities for the elderly. Subsequent phases of the project included studies of port workers, political organisations, factories, Arab villages, prisoners, slum dwellers and old people's homes.[35] Given the diversity of the subjects studied it is not surprising that the overall focus of the project shifted from immigration to 'the relations between the state and its citizen ... [and] the nature of centralized bureaucratic organisation'.[36] All of the researchers published studies based on their fieldwork – a successful publication rate on a par with the RLI/Central African research for which the department is best known.

In Israel, the Bernstein scheme led to a flourishing of social anthropology. Tel Aviv University had established its own Department of Social Anthropology and Sociology in the early 1960s. The project would provide the nucleus of the new department, with Emmanuel Marx, Shlomo Deshen and Moshe Shokeid all taking up posts there upon the completion of their contracts with Manchester.[37] The project was due to conclude with the production of a general report summarising the research undertaken under its aegis. To this end, in late 1974, Gluckman went to the Hebrew

University of Jerusalem as Lady Davis Distinguished Visiting Scholar; he was due to stay for six months.[38] However, he died suddenly in Jerusalem on 13 April 1975 as a result of a heart condition.

The Bernstein project would be the final achievement of the Manchester department in its original incarnation under Gluckman's direction. It represented many of the ideas and practices that had characterised the department in its quarter-century of existence. It was a team research project that balanced intellectual questions (what does modernity look like?) with practical applications (how best to integrate immigrants?). It drew comparisons across shared themes ('the village' or 'the role of conflict in creating and maintaining norms'). In this, it reflected the Manchester Department of Social Anthropology and Sociology's rich heritage of innovative work that interrogated assumptions about modernity and community life, work that was deeply embedded in its local area, but that also looked outward at an ever-changing world.

Notes

1 For example, A. L. Epstein, *Politics in an Urban African Community* (Manchester: Rhodes-Livingstone Institute and Manchester University Press, 1958); J. Clyde Mitchell, *The Yao Village: A Study in the Social Structure of a Nyasaland Tribe* (Manchester: Manchester University Press, 1956).

2 Mary Douglas, '270: reviewed work "Tribal Cohesion in a Money Economy" by William Watson', *Man*, 59 (1959), 168.

3 Works addressing the 'Manchester School' include Robert J. Gordon, *The Enigma of Max Gluckman: The Ethnographic Life of a 'Luckyman' in Africa* (Lincoln: University of Nebraska Press, 2018); David Mills, *Difficult Folk? A Political History of Social Anthropology* (New York: Berghahn Books, 2008); Lyn Schumaker, *Africanizing Anthropology: Fieldwork, Networks, and the Making of Cultural Knowledge in Central Africa* (Durham, NC: Duke University Press, 2001); and Richard Werbner, *Anthropology After Gluckman: The Manchester School, Colonial and Postcolonial Transformations* (Manchester: Manchester University Press, 2020).

4 UML, VCA 7/87, Particulars of Appointment of Reader in Social Anthropology, 1949.

5 UML, VCA 7/87, letter from Raymond Firth to Arthur Lewis, n.d.

6 UML, VCA 7/357, 'Readership in Social Anthropology', 27 April 1949. David Mills attributes the transition from readership to chair to successful negotiation on the part of Gluckman; Mills, *Difficult Folk?*, p. 101.

7 Bodleian Library, James Clyde Mitchell Papers (hereafter JCM), Box 5, Folder 1, Item 44, letter from Max Gluckman to Clyde Mitchell, 6 May 1949.
8 Bodleian Library, JCM, Box 5, Folder 1, Item 139, letter from Max Gluckman to Clyde Mitchell, 11 December 1950.
9 Schumaker, *Africanizing Anthropology*, p. 110.
10 LSE Archives, Association of Social Anthropology Papers, ASA, A10, letter from Max Gluckman to Ioan Lewis, 20 July 1965.
11 Bodleian Library, JCM, Box 5, Folder 1, Item 139, letter from Max Gluckman to Clyde Mitchell, 11 December 1950; Bodleian Library, JCM, Box 4, Folder 1, Item 39, letter from John Barnes to Clyde Mitchell, 4 September 1949.
12 Royal Anthropological Institute Archives (hereafter RAI), Gluckman Papers, Folder 'RJF', Item 233, letter from Max Gluckman to Ronnie Frankenberg, 1 March 1957. Nothing came of this publishing plan. Note too that Gluckman's reference to the 'Manchester School' in this letter was two years before Mary Douglas apparently coined the phrase in print.
13 UML, VCA 7/404 Folder 1, correspondence relating to Frankenberg visa ban, 1953. Frankenberg gives his own account of these events in '*Village on the Border*: a text revisited', in Ronald Frankenberg, *Village on the Border: A Social Study of Religion, Politics and Football in a North Wales Community* (1957) (Prospect Heights, IL: Waveland, 1990).
14 The National Archives, CO 1035/135, 'Brief for the Minister', February 1953.
15 RAI, Gluckman Papers, Folder 'RJF', Item 117.1, 'Draft Scheme', March 1953.
16 Frankenberg, *Village on the Border*.
17 Derek Allcorn, 'The Social Development of Young Men in an English Industrial Suburb', PhD thesis, University of Manchester, 1955.
18 Allcorn did publish an article about his research, 'The unnoticed generation', *Universities and Left Review*, 4 (1958), 54–8.
19 Frankenberg, '*Village on the Border*: a text revisited', p. 179.
20 RAI, Gluckman Papers, Folder 'RJF', Item 220, letter from Max Gluckman to Mr Charlton, 30 October 1956. Reasons for the rejection were apparently that it lacked statistics on unemployment and devoted too much space to descriptions of committee meetings. The book would be published by Cohen & West instead.
21 Rockefeller Archive Center, FA733, Reel 1009, C-344, letter from Max Gluckman to the Ford Foundation, 25 July 1957.
22 Ibid.
23 UML, VCA 7/404. Folder 3, letter from Leslie Farrar-Brown to William Mansfield Cooper, 17 March 1958.
24 Watson would not ultimately publish any of his research, with the exception of an article: William Watson, 'Social mobility and social class in industrial communities', in Max Gluckman (ed.), *Closed Systems and Open Minds* (Chicago: Aldine, 1964), pp. 129–57.
25 Richard Gillespie, *Manufacturing Knowledge: A History of the Hawthorne Experiments* (Cambridge: Cambridge University Press, 1991).

26 Tom Lupton, *On the Shop Floor: Two Studies of Workshop Organization and Output* (Oxford: Pergamon, 1963); Sheila Cunnison, *Wages and Work Allocation: A Study of Social Relations in a Garment Workshop* (London: Tavistock Publications, 1966).

27 Ronald Frankenberg, 'A social anthropology for Britain?', in Ronald Frankenberg (ed.), *Custom and Conflict in British Society* (Manchester: Manchester University Press, 1982), p. 26.

28 LSE, ASA, A1.3, 'The Teaching of Social Anthropology', 22–25 September 1958.

29 RAI, Gluckman Papers, 'Personal and Manchester University Business', letter from Emrys Peters to Ely Devons, 12 November 1958.

30 UML, VCA 7/404, Folder 6, letter from Max Gluckman to William Mansfield Cooper, 8 December 1960.

31 For an example of this relating to Manchester anthropologists, see Geoffrey Gray, '"In my file, I am two different people": Max Gluckman and A.L. Epstein, the Australian National University, and Australian Security Intelligence Organisation, 1958–60', *Cold War History*, 20:1 (2020), 59–76, doi: 10.1080/14682745.2019.1575367.

32 Relatively little has been written on this scheme. Two articles by participants in the scheme provide an overview of its scope and impact: Emmanuel Marx, 'Anthropological studies in a centralized state: Max Gluckman and the Bernstein Israel Research Project', *Jewish Journal of Sociology*, 17:2 (1975), 131–50; Moshe Shokeid, 'Max Gluckman and the making of Israeli anthropology', *Ethnos*, 69:3 (2004), 387–410.

33 UML, VCA 7/949, Folder 2, letter from Max Gluckman to William Mansfield Cooper, 3 July 1961.

34 RAI, Gluckman Papers, Personal and Manchester University Business, 'Report on Work Being Carried Out in Middle Eastern and Arab Countries in Faculty of Economic and Social Studies', n.d. [1966].

35 Marx, 'Anthropological studies', 134–5.

36 Ibid., 148.

37 University of East Anglia Archives, Solly Zuckerman Papers, SZ/BIRT/1, 'Minutes of a Meeting of the Bernstein Israeli Research Scheme', 24 September 1968.

38 UML, VCA 7/949, Folder 1, letter from Max Gluckman to A. L. Armitage, 14 August 1974.

Shock city: Michel Butor and W. G. Sebald in Manchester

Catherine Annabel

Manchester's University launched many academic careers. But for two pivotal figures in late twentieth-century European literature, Michel Butor and W. G. Sebald, the experience of the University and the city also shaped their literary careers, and resonates through their work. Butor is somewhat neglected now, but played a key role in what was known as the *nouveau roman,* and wrote prolifically in various genres. Most importantly in this context, he is the author of one of the great Manchester novels, *L'Emploi du temps (Passing Time).*[1] Sebald's career was much briefer, cut short by his sudden death in 2001, but he remains both widely read and highly regarded.

Both writers came to Manchester University as language assistants, Butor to the French Department in 1951, and Sebald to German in 1966. While the University gave them both the experience to move on in academia (Butor to Geneva University, Sebald to the University of East Anglia), the city itself was just as crucial in their future literary success.

For both, Manchester was a shock. Butor arrived in autumn 1951, fresh from teaching in Egypt, where he had been soaked in sun, as he put it, and the climate appalled and depressed him.[2] Sebald's upbringing in rural Bavaria had done nothing to prepare him for the archetypal industrial city, and the grime and dereliction evident in parts of the city affected his mood. Both spent much time on solitary and melancholic walks around the city.

While in Manchester, Butor began writing his first novel, *Passage de Milan*.[3] But it was his second that captured the experience of Manchester, renamed as Bleston, in a remarkable diary novel where rain and fog permeate every page, and the city streets are a labyrinth to trap the unwary. In *L'Emploi du temps*, the city is not only a physical reality but a powerful presence, a sorcerer, a Hydra, a Minotaur. Nonetheless, Manchester is recognisable, not just through the weather, but through many of its landmarks. The opening pages of the book initially suggest a prosaic account, laced with wry humour, of an inauspicious arrival in a grim northern town, where the food is as awful as the weather and the welcome as chilly. But the book's time scheme rapidly becomes more complex, as the diarist's state of mind becomes more troubled, and the mundane events described are at odds with the intensity of the diarist's emotions, with a sense of peril, threat and incipient violence. Butor himself suggests a link between Bleston and Paris, which might seem improbable, unless one considers the Paris of Butor's adolescence during the Nazi Occupation, when everyday life was full of menace, and darkness and fog pervaded the city of lights.[4] This is a complex and fascinating work, which challenges the reader to navigate the narrative, as it becomes steadily more labyrinthine, and which readers and critics have interpreted in very different ways since its publication.

When W. G. Sebald arrived in Manchester in autumn 1966, after a period at Freiburg University in Switzerland, during a difficult settling-in period (like Butor, he struggled to find decent lodgings, struggled with the language, and with the new environment), he was reading *L'Emploi du temps*. His copy of the novel is heavily annotated, with vigorous underlinings, exclamation marks and other marginal comments, making his identification with the diarist clear.[5] Arguably, reading Butor's account contributed to Sebald's melancholy – certainly, the mood of the book chimed with his own.[6]

Sebald's poem 'Bleston: A Mancunian Cantical', dated January 1967, interweaves quotations from and allusions to *L'Emploi du temps* with copious intertextual references, in a way that is familiar from his later prose works and his poem *After Nature*, which also

V4.1 Michel Butor's *Passing Time* (Jupiter Books, 1965) was first published in French in 1956 as *L'Emploi du temps*. © Alma Books (www.almabooks.com). Photograph by Patrick Tookey.

includes a section relating to Manchester, its history and climate, and its Jewish community.[7] In turn, this fed into the Max Ferber section of *The Emigrants*.[8] Sebald's encounter with his Jewish landlord Peter Jordan, who had left Nazi Germany in 1938 and whose parents were killed at Kaunas Fort in 1941, had a profound influence on his writing and his engagement with the Holocaust.[9] He had seen

V4.2 University of Manchester c. 1970, shortly after W. G. Sebald left. The buildings were still blackened with soot; they were cleaned around 1972. Courtesy of MMU Visual Resource Centre.

the films of the liberation of Dachau, and he had closely followed the progress of the Auschwitz trials. But he encountered here for the first time survivors of those horrors, and he talked for hours to Jordan during that early period and in subsequent years, using his history and that of his family for his account of Max Ferber.

For both Butor and Sebald, the encounter with Manchester, the shock city of the nineteenth century, was a catalyst.[10] The portrait they paint of the city is both vividly recognisable and unreliable, as they see it through the lens of the melancholy that Manchester's darkness, fog and grime provoked. But this encounter also suggested connections with their own traumatic experiences, and those of others, connections that enriched their writing.

Notes

1 Michel Butor, *L'Emploi du temps* (Paris: Editions de Minuit, 1956); *Passing Time*, trans. Jean Stewart (Manchester: Pariah Press, 2021).
2 Georges Charbonnier, *Entretiens avec Michel Butor* (Paris: Gallimard, 1967), p. 96.
3 Michel Butor, *Passage de Milan* (Paris: Editions de Minuit, 1954).
4 Catherine Annabel, '"Perdu dans ces filaments": Labyrinths and Intertextuality in Michel Butor and W. G. Sebald', PhD thesis, University of Sheffield, 2022, Annabel, Catherine 130127477 final.pdf (whiterose.ac.uk) (accessed 28 March 2024).
5 Sebald's annotated copy is held in the Deutschesliteratur Archiv, Marburg.
6 Richard Sheppard, 'The Sternheim years: W. G. Sebald's *Lehrjahre* and *Theatralische Sendung* 1963–75', in Jo Catling and Richard Hibbitt (eds), *Saturn's Moons: W. G. Sebald – A Handbook* (Oxford: Legenda, 2011), pp. 64–81.
7 W. G. Sebald, 'Bleston: a Mancunian cantical', in *Across the Land and the Water*, trans. Iain Galbraith (London: Hamish Hamilton, 2011), pp. 18–22; W. G. Sebald, *After Nature*, trans. Michael Hamburger (London: Hamish Hamilton, 2002).
8 W. G. Sebald, *The Emigrants*, trans. Michael Hulse (London: Harvill Press, 1996).
9 Thomas Honickel, *Curriculum Vitae: Die W. G. Sebald-Interviews*, ed. Uwe Schütte and Kay Wolfinger (Würzburg: Königshausen & Neumann, 2021), pp. 105–14.
10 Asa Briggs, *Victorian Cities* (London: Penguin, 1990), p. 96.

13

Dorothy Emmet: 'for administrators whose hearts are with the anarchists, and anarchists who can have a heart for the administrators'

Rachael Wiseman

In 1961 Dorothy Emmet, head of the University of Manchester's Philosophy Department and Manchester's only female professor, was told that she was next in line to be faculty dean. She was 'not keen'.[1] She had recently returned from a research trip to Columbia University, and a series of conversations with the American sociologist Robert K. Merton had seeded an idea for her next book. That idea had since passed her 'bus and bath test' – that is, she found its subject-matter 'coming out on top' at such moments of idle solitude – and she was looking forward to giving it more serious thought.[2] Furthermore, she took no delight in the prospect of an administrative role at a time of growing student unrest, when '"soul-less" had become an almost inseparable epithet of "bureaucrat" as applied to those running the universities'.[3] In Manchester the Students' Union was becoming increasingly political, and just months earlier its newspaper had been closed down by the vice-chancellor following a series of 'scurrilous stories', including one 'attacking the University Bursar'.[4] However, Emmet thought she would make the best of it, writing to Merton: 'One consolation in having to do a spell in administration is that if approached with an interest in Sociology and Ethics, it may well provide some fieldwork, though not necessarily such as would be describable in public.'[5]

Two years later, she emerged from her service somewhat triumphant:

Dear Bob, this is to tell you that I'm still alive ... [Being dean] has left precious little time for writing. however, it hasn't been a waste of time from the "Soc. & Ethics" point of view, as I can feel a number of problems more on the pulses after having to do a fairly heavy administrative job for two years, with the pressure it involves.[6]

In May 1965 she submitted the manuscript of *Rules, Roles and Relations* to her publisher, writing in the preface that she had written the book for the 'administrators whose hearts are with the anarchists, and anarchists who can have a heart for the administrators'.[7]

Rules, Roles and Relations would be Emmet's last major work while at Manchester, and its publication marked the end of two decades of collaboration and conversation between Emmet and a remarkable cast of social scientists, political and management theorists, anthropologists and philosophers, centred on the University's Department of Social Anthropology. For Emmet, those conversations inspired a pair of books at the intersection of these disciplines. This chapter tells the story of their creation by situating them in the broader context of Emmet's life, political and religious outlook, and distinctive metaphysical vision. I hope it makes the case for renewed scholarly engagement with Emmet's philosophy – almost entirely neglected by current writers[8] – and raises in a new way old questions about the role and social function of the academic.

Introducing Dorothy Emmet

Dorothy Emmet was the eldest daughter of Gertrude Julia Weir and Cyril Emmet. In her memoir, Emmet writes that her father 'belonged to what is now probably an extinct species, the scholar country parson'.[9] His church was home to swallows and bees (the latter once responsible for showering the congregation with honey),[10] and the vast and the unmodernised vicarage (without central heating or plumbing) was surrounded by meadow and orchard.[11] Emmet and her sister were first 'taught at home by a not very intelligent governess'; then, at the age of 14, she enrolled at St Mary's Hall, Brighton – an 'austere' school for the daughters of clergy. Its sister school, Cowen Bridge, was attended by the Brontë sisters and appears as Lowood School in Charlotte's *Jane Eyre*.[12]

In 1923, at the age of 19, Emmet went up to Lady Margaret Hall, Oxford, to study 'Mods and Greats' – a combination of classics, ancient history and philosophy. She arrived just three years after the University had permitted the first women to take degrees (and twenty-five years before they would be allowed to do so in Cambridge), and at a time when many in the country found something absurd, if not unnatural, in the idea of a woman playing the role of scholar.[13] As Lady Margaret Hall did not have any philosophy tutors (the women's colleges were kept small and poor by the University's 'Limitation Statute' and could afford to employ only a handful of scholars), Emmet took her philosophy tutorials at Balliol with the moral and political philosopher A. D. (Sandy) Lindsay.[14] She found in him a tutor who was 'full of good sense' and 'concerned that [his students] should be given the truth', rather than rigorously trained in the art of winning arguments.[15] She would later describe herself as a 'Balliol stepdaughter'.[16] In 1926 Emmet had something resembling 'a conversion experience' at a lecture by R. H. Tawney in support of the General Strike, and this was the beginning of a lasting and serious concern with politics.[17] She taught for the Workers' Educational Association, spending her summers in the Rhondda Valley reading Plato's *Republic* with unemployed miners. Many years later she almost fell out with Karl Popper over his claim (in *The Open Society and Its Enemies*) that teaching Plato's text turned students into fascists. She was not impressed with his explanation that he had 'put things in an extreme way [so that] people would believe at any rate part of what he was saying'; 'As he had only recently come to the country I took the liberty of telling him that I did not think he would find that this worked in England.'[18]

After graduating with a First, Emmet applied for, and was awarded, a Commonwealth Fellowship to study with A. N. Whitehead, a distinguished professor regarded by many as 'perhaps the most original philosophical genius' at the time.[19] In her final year at Oxford, Emmet had read his *Science and the Modern World* (1932), the first of his three major works developing 'process metaphysics'

– a cosmology according to which our fundamental intuition with respect to reality is not that *something is the case*, but that *something is going on*.[20] Whitehead once remarked to Bertrand Russell (whose outlook was very much of the former kind): 'you think that the world is what it looks like in fair weather at noon-day. I think it is what it seems like in early morning when one first wakes from deep sleep.'[21] Emmet was both attracted to and bemused by Whitehead's worldview, and though she initially found herself more inclined to Russell's sunny civilisation than Whitehead's disorienting hinterland, she was determined to understand him.[22] Whitehead had recently emigrated to America and was teaching at Harvard, but as this was a male-only institution Emmet enrolled at its sister college, Radcliffe. This proved to be a blessing: at Harvard Whitehead lectured to 70 young men; at Radcliffe, Emmet was one of eight.[23] She and her fellow classmates were completely smitten with Whitehead – she described him to her mother as 'a perfectly wonderful old man, of the most seraphic disposition'.[24] On her return to England, she spent a research fellowship at Somerville College, Oxford, writing *Whitehead's Philosophy of Organism* (1932), the first introduction to his metaphysics. She found a home in Whitehead's hinterland and his vision of the world as a myriad of interacting processes sustained by continuous creative activity would remain a touchstone for her thought. As we will see, Whitehead's move away from a fundamentally static picture of nature and reality to one in which *creative force* and *process* are foundational categories would later find expression in Emmet's approach to the social world.

Emmet's first lectureship was at King's College in Newcastle, then part of Durham University. The department was composed of a lone professor and herself (a 'man and his dog and ... I was the dog', she joked).[25] In 1938 she moved to the similarly tiny Manchester Philosophy Department as a lecturer in philosophy of religion. It was there, during the Second World War, that she wrote her own work of metaphysics: *The Nature of Metaphysical Thinking*. At the end of the war she was made head of department and in 1946 became Manchester University's lone female professor, and only the second female philosophy professor in Britain (Susan Stebbing being the first). The *News Chronicle* reassured its readers that the promotion and the writing of 'several philosophy books' did not interfere with her more natural domestic duties: 'Professor in the Kitchen: She cooks for relaxation' declared the headline, alongside a photo of a smiling Emmet in a housecoat.[26] Emmet would remain at Manchester until her retirement in 1966, overseeing the growth of the Philosophy Department from 'a handful of students and one lecturer' to a community of many academics and 'over 400 students'.[27]

The final decades of her life were extraordinarily productive. She retired to Cambridge to live with her lifelong friends, the philosophers Margaret Masterman and her husband Richard Braithwaite. Their home became the base camp for the activities of the Epiphany philosophers – a community of religiously minded philosophers and scientists that included at one time Mary Hesse, Anthony Appiah and Rowan Williams. Emmet admired and corresponded warmly with Iris Murdoch,[28] and published four further books on metaphysics and ethics, as well as a philosophical memoir which includes cameos from (among others) Gandhi, Archbishop William Temple, Arthur Koestler, Ludwig Wittgenstein, Karl Popper, Iris Murdoch and Alan Turing.[29]

The nature of metaphysical thinking

To appreciate the character of the philosophical vision that Emmet brought to the circle of social scientists, political and management theorists and anthropologists that she was to encounter in the Manchester of the 1950s, we must first join her during the Second World War as a newly appointed lecturer. Emmet wrote her second book, *The Nature of Metaphysical Thinking*, while scanning the skies above Manchester for German bombing raids. Though excluded from conscription on the grounds of her occupation, she was, like others in her position, required to do national service alongside her day job; and so for much of the war she spent one night in six fire-watching. In the later years of the war bombs were rare, so this meant 'long silent hours in an almost empty building'.[30]

PROFESSOR IN THE KITCHEN

She cooks for relaxation

News Chronicle Reporter

MISS Dorothy Emmet, M.A. (Oxford) whose appointment as Professor of Philosophy at Manchester University was announced yesterday, believes that cooking gives relaxation from hard thinking.

Miss Emmet is the only woman professor of philosophy in this country.

When I saw her last night she was preparing a meal at her house in Yew Tree-lane, Wythenshawe. She runs the house alone, with occasional help, and enjoys gardening.

She came to Manchester University from King's College, New-castle-on-Tyne, in 1938, as lecturer in the philosophy of religion.

Leaving Lady Margaret Hall, Oxford, with a first in "Greats," Miss Emmet went later to Harvard University as a Commonwealth scholar. She has written several books on philosophy.

13.1 *News Chronicle* report of Dorothy Emmet's appointment as Professor of Philosophy at Manchester. 'She runs the house alone, with occasional help, and enjoys gardening.'

Emmet had arrived at Manchester with a mission. Accepting the position, she had written to the vice-chancellor that 'in these present times when we see the natural cohesive processes of society breaking down it is essential to be thinking more deeply about the questions with which the philosophy of religion is concerned'.[31] She would later recall the 1930s as 'the unhappiest time of [her] life', not personally, but 'in respect of what was going on in the world, and the chilling uncertainty about what could be done about it'.[32] This including rising unemployment at home and the desperate situation in Europe, which made Jewish refugees (including many academics) a familiar sight in both Oxford and Manchester. Emmet held that in philosophy 'one is where one is, and must try to see what one can from there',[33] and her sense of the social and political urgency of questions about religious life reflects her commitment to setting out from where she found herself, however inhospitable that place was. Her feel for the possibility of a tripartite religious, political and philosophical vocation had likely been shaped by her encounters with Lindsay and Whitehead, but also by her conversations with William Temple, to whom she had given tutorials on Whitehead's metaphysics as he prepared his Gifford Lectures.[34] Emmet saw the need for 'some metaphysical link' between the teachings of philosophy and those of religion. Thus, she found she 'had to look at what, in principle … [s]he thought about metaphysics'.[35]

Emmet's decision in 1938 to 'think about metaphysics' was, to put it mildly, not a fashionable one. More specifically, metaphysics in Britain was in the middle of a methodological crisis.[36] 1936 had seen the publication of A. J. Ayer's wildly popular and absurdly polemical *Language, Truth and Logic*. According to Ayer, because the metaphysician's assertions could not be verified (or falsified) by experience they were not genuine statements of fact: they were (at worse) nonsense or (at best) poetic utterances.[37] This positivist manifesto – aimed at the British Idealists and Oxford Realists who had been his (and Emmet's) lecturers[38] – had turned 'metaphysician' into an insult and (as Mary Midgley recalled) altered the use of 'the term "metaphysic"' so that it meant 'something outside the range of rational thought'.[39]

To add to the difficulty of her task, Emmet did not mean to revive or repurpose Idealism or Realism but rather to bring about a new period of metaphysical thinking, one that did not suffer from the fatal flaws of its predecessors. The Idealists, she thought, began well, from a pair of undeniable facts: first, that 'our primary awareness is already an ordering and interpretive activity'; second, that it is an 'impossibility' to go 'beyond experience and stat[e] in any intelligible fashion the relation between experience and anything other than experience'.[40] The Realists put these truths beyond reach when they claimed that 'knowing in no way alters or modifies the thing known'.[41] Emmet thought this 'as certainly wrong as anything in philosophy can be'.[42] As an undergraduate she had sat through lectures in moral philosophy by the Realists W. D. Ross and H. A. Prichard, astonished by their confidence that they lived in an unchanging world, waiting to be known and blandly indifferent to the activities of its human inhabitants.[43] But the Idealists, Emmet thought, spoiled their point by insisting that they could discern a 'necessary systematic character within experience',[44] one that revealed the world to be 'a great construction of some absolute mind'.[45]

Emmet located herself 'at the end of a period of metaphysical thinking' and stated an ambition, framed in terms that self-consciously echo Kant, to inaugurate a 'new constructive movement of metaphysics'.[46] Her starting point is the two facts that the Idealists recognised – that 'our primary awareness is already an ordering and interpretive activity' and that we cannot say anything about 'the relation between experience and anything other than experience' – but she rejects their vision of a vast and interconnected rational structure. Now at home in Whitehead's hinterland, Emmet insists that we 'start from experience as a vague and confused world of ideas', with 'sense data ... the most elementary' component.[47] From this starting point, we face the task of actively selecting data that are important, seeking out relations between them, and so 'translat[ing] a multiplicity of data into some definite form'.[48] This activity, Emmet thinks, is not (as the Idealists claimed) a matter of intuiting already existing structural forms, but is rather a creative and personal endeavour, in which judgements about what is real and what is important are interwoven. Because each individual is where she is, and has only a partial and limited view, the 'process of selection, abstraction, interpretation'[49] will necessarily involve creative 'leaps' – novel projections, metaphors, myths and analogies will be needed to bring discrete data into relation. Armed with this account of experience, Emmet's bold claim is that a 'new constructive movement of metaphysics' should cast the metaphysician as an individual working at the extreme limit of this integrative process. She (the metaphysician) is someone whose desire to bring her experience into 'definite form' leads her to hazard a 'leap' from those 'resemblances [that] are sufficiently strong to seem significant' and 'to extend the analogy still further to say something about the nature of reality' as a whole.[50] For Plato, the dominant analogy was from mathematics; for Aristotle, from biology; for Hegel, from 'the movement of a sustained philosophical argument'.[51]

For Emmet, then, a metaphysical system is not 'a kind of suprascience'[52] but is rather 'a man's personal reaction to the world' – it is his (or her) attempt to fashion a response to that *something* that is *going on*, and in the midst of which he has awoken. Emmet compares the metaphysician to an artist insofar as both seek 'precision in expressing his response to experience', and to a literary critic insofar he judges after reflection.[53] In each case, what is aimed at is a 'total judgment' – one that transcends particular assertions about this or that feature and hazards to be a response to the thing as a whole.[54] As each metaphysician is wherever he finds himself, with only a partial and provisional view of what is *going on*, so too 'each metaphysical view will be partial and provisional'.[55] Nevertheless, just as an artist seeks to convey a truth about what is real and about what is important, so too does a metaphysician – this prevents the metaphysician's vision from being *merely* his subjective response. And just as the vanity of an artist can ruin her work, so too a metaphysician must beware the temptation of 'egoistic attempts to dominate the world in theory in terms of a merely personal reaction'.[56]

The book's publication coincided with the Philosophy Department's search for a new chair, and it seems likely that she would not have been shortlisted without it – her friend, Michael Polanyi, took the trouble of writing to the vice-chancellor: '[Miss Emmet] has just published a book on "The Nature of Metaphysical Thinking" … and I suppose this may considerably affect the valuation of her merits'.[57] The book announced her transition from a highly respected interpreter of Whitehead to a metaphysician in her own right. Letters of support followed from a number of distinguished professors: John W. Harvey did 'not think [he] could suggest anyone else whose claims and qualifications are comparable to those of Miss Emmet';[58] G. C. Field thought her new book 'distinctly impressive' and found her 'thoroughly worthy of a Chair'.[59] However, the vice-chancellor eventually decided that there was no suitable candidate for the role, and Emmet was instead promoted to reader and made head of department. We cannot know how much weight the search committee gave to Ross's assessment of Emmet as the least able of a shortlist of six.[60] However, she did not wait long, and her promotion to a professorship came the following year.

Getting in on the act with the practitioners

Though Emmet's interest in metaphysical thinking put her at odds with the positivistic tendency of linguistic philosophy in its early years, as the movement matured into a recommendation to philosophers that they pay careful attention to the way in which words are used Emmet came to 'nearer to appreciating' it. Indeed, in this new guise, it connected to the 'sharpen[ed] feeling for the meaning of words' that was a natural by-product of her training in Classics.[61] In 1946, in a paper on 'The Idea of Importance' for the Aristotelian Society, she tried her hand at the method, showing (as was already implicit in her book) that there is a use of the word 'important' that ties it closely to 'real'.[62] In Oxford, so-called 'ordinary language philosophers' Gilbert Ryle and J. L. Austin largely stuck to using the method as a prophylactic against metaphysics (by showing that speculative theorists of the past had been confused about the meaning of their own words) and in skirmishes with other philosophers.[63] (Mary Warnock, a student immediately after the war, recalls the delight with which the men brandished their new weapon: 'I don't understand.')[64] But Emmet felt that that there was more profit to be had in looking outside philosophy and finding 'some subject matter which throws up linguistic and conceptual problems and where philosophers might be able to get in on the act with the practitioners'.[65] And as the 1950s began, Emmet found herself perfectly placed to do so by integrating herself into Manchester's newly established Department of Social Anthropology. Emmet began attending their interdisciplinary seminars, which were filled with terms calling for linguistic analysis: structure, purpose, intention, rule, society, practice, explanation, meaning. She was welcomed and soon established as the 'friendly neighbourhood philosopher'.[66]

The extreme fecundity of the collective that emerged out of those seminars is astonishing – here is one of those moments in academic life when space, time and circumstance deliver up a set of perfectly matched conversationalists, all at the height of their intellectual and creative powers. Emmet formed a loose 'quartet' with anthropologist Max Gluckman, political theorist William (Bill) MacKenzie and economist Ely Devons; a steady flow of books followed that 'read in part like a series of reflections on our discussions'.[67] Michael Polanyi was a regular attender; Alasdair MacIntyre and Victor Turner joined as graduate students. The group invited eminent speakers – including Robert K. Merton, Jean Piaget and Alfred Radcliffe-Brown – and in 1949 hosted a discussion on 'Mind and the Computing Machine' with Alan Turing.[68] Seminars spilled over into 'various environments, some of them not very academic'[69] – these included the College Arms and the football terraces at Old Trafford. Emmet was taught to 'drink beer in pubs'. She later described the 1950s as 'the best time I have known for doing philosophy in an interdisciplinary setting'.[70] For MacKenzie the University was 'a place of intellect and friendship'.[71]

In Emmet's metaphysics, the dominant image is that of a creative individual who awakes in the midst of interacting and unfolding processes that she must, from her limited perspective, seek to order

13.2 The College Hotel (or 'College Arms'), 1958. Emmet recalled lively interdisciplinary exchange at this centrepiece of the academic community. It was here that she was taught to 'drink beer in pubs'. Courtesy of Manchester Libraries, Information and Archives, M49426.

and, ultimately, to influence. This vision is heady in its metaphysical register but, learning from the work of her new friends, she transposed it (in *Function, Purpose and Power* and *Roles, Rules and Relations*) into a vision of social and institutional life so practically grounded that one reviewer bemoaned her 'slump' back to earth,[72] and Geoffrey Vickers – head of recruitment and training for the National Coal Board – immediately recognised its truth to managerial life.[73]

Function, Purpose and Power is, on the surface, a sympathetic critique of functionalism as a method in sociology.[74] According to that approach, which was dominant among social scientists at the time, sociological explanation is 'explanation of facts by the part they play within the integral system of culture, by the manner in which they are related to each other within the system, and by the manner in which this system is related to the physical surroundings'.[75] A social scientist does not, that is, explain why Jack and Jill marry by speaking of Jack's intentions and Jill's feelings, but by describing the role of marriage in sustaining the affective and economic ties that are crucial to the maintenance of the social structure. Functionalism, then, is at root an analogical extension of biological and organic concepts into the realm of social explanation; the homeostatic telos of an organism becomes the self-maintaining telos of a culture.[76] But for Emmet, this extension – aside from its clear failure to answer (or indeed ask) regarding the value of self-maintenance – obscured from view the individual human beings who, at a particular moment in time, make up a society. Emmet thought that if social scientists lost from view the *individuals* who performed the roles, they would leave themselves unable to explain the politically essential process of social change. Each, for example, teacher *in concreto* is an individual, who performs her role in her own, personal way. She has the power to ask herself what it means, to her, to be a teacher, and how she as an individual can inhabit the role, and whether the social needs that teaching meets might be served better if the role defined for her was different. Emmet's favoured example was that of Florence Nightingale – she appears in both books – who did not just inhabit the role of 'nurse', but rather laid down 'a new pattern' when she 'created the role of the trained professional nurse', transforming it from 'the squalid association of Sairy Gamp' into a profession with high standards and institutional recognition.[77] An individual, in the right setting, has the creative potential to change a role she inhabits and the social structure in which the role is embedded.

Emmet proposes a new dominant analogy for sociologists: the social world is not a homeostatic organism in self-sustaining equilibrium, but an ever-changing and surprising 'theatre of activities', in which the creative activity of persons-in-roles both keeps the show *going on* and, at the same time, contains the possibility of improvisation and novelty – of going off-script. Activity within an unfolding cultural, administrative or institutional process can variously 'form, dissolve, re-form, and sometimes produce a new kind with capabilities for new kinds of activity'.[78]

Anarchists and bureaucrats

The 1950 and 1960s were a time of great social change, and in philosophy as well as in society at large an anarchic temperament was gaining ground. This tendency made authenticity, sincerity and (as we might put it today) 'living your truth' the central (perhaps *sole*) virtues of moral and political life. In Sartre's famous example, to which Emmet devotes a long discussion in *Rules, Roles and Relations*, all role playing is seen as *mauvaise foi*. The waiter 'is playing at being a waiter in a café', and in doing so is acceding to 'society['s] demand that he limit himself to his function ... just as the soldier at attention makes himself a soldier-thing'.[79] For the existentialist, purity of purpose means acting outside of role and convention, and forming intentions and making plans in light only of one's authentic desires and freely chosen principles.[80] Emmet thought that the existentialist was right to issue a warning: 'there is a real temptation to seek security in having a function, especially if it is one that carries "status"'.[81] But she could not agree that what individuals in a society demand of each other is total absorption into a function nor that an anarchistic life outside social roles would be desirable or possible – *anarchist rebel* is, after all, a role of its own.

Sadly for us, Emmet's fieldwork as dean did not result in an anthropological study of *Custom and Conflict in a Northern University*[82] – as she predicted, her findings were 'not such as would be describable in public'. But she returned from her ethnological study of the higher bureaucratic structures of the University with a heart for both the anarchists and the bureaucrats, and a clearer sense of the need for symbiosis between them. The anarchists, she knew, depended on the bureaucrats. For though 'how original a person dares to be in carrying out a role may largely depend on how strong a character he is', how far he can go 'and still be tolerated and not ostracised as a "deviant"' is not up to him. Rather it will 'depend on the attitude of other people in the society to other strong

characters, and how much concern they have for encouraging possibilities of vocation'.[83] Here the anarchist might appreciate the
help offered by a wise administrator who creates space within their
institution's roles for 'the development of creative originality' that
allows Nightingale-like figures to emerge.[84] Equally, the anarchist
can find a place among the bureaucrats. A wise dean who refuses
the comfort of the epithet 'soul-less' and consults her own anarchistic
heart can offer protection to those rebels who would rather improvise
than follow the script they are offered. If a university is not the
sort of institution in which such experiments can be allowed to
unfold, it is hard to see where they might. This suggests that a
defence of universities – their institutional autonomy and the freedom
of the individuals that make them up – might also mention their
importance as 'theatres of activity' where wise deans extend toleration
to anarchists and creative individuals find institutional structures
within which they can experiment with new patterns for learning,
working and living.

* * *

My biggest thanks are (as always) to Clare Mac Cumhaill, whose
collaboration, philosophical conversation and friendship continues
to be invaluable. Thanks too to Laurence Blum, Vic Seidler and
Mark Hopwood for all their brilliant (and, no doubt, here mangled)
insights on Emmet as part of our reading group. I am grateful to
Stuart Jones for the opportunity to write about Emmet, and to Clare,
Larry, Vic and Stuart for comments on a draft.

Notes

1 Rare Book & Manuscript Library, Columbia University in the City of New
   York, Robert K. Merton papers, Correspondence with Dorothy Emmet
   (henceforth CDE), letter from Dorothy Emmet [DE] to Robert K. Merton
   [RKM], 19 October 1961,
2 Ibid.
3 Dorothy Emmet, *Philosophers and Friends* (London: Macmillan, 1996), p.
   94. Though the peak of student activism in both the US and Britain was
   1968, British student unions were increasingly political through the 1960s.

See Caroline Hoefferle, *British Student Activism in the Long Sixties* (New York: Routledge, 2013), esp. chs 1 and 2.

4 Brian Pullan, *A Portrait of the University of Manchester* (London: Third Millennium, 2007), pp. 166, 169.

5 Letter from DE to RKM, 19 October 1961, CDE.

6 Letter from DE to RKM, 11 October 1961, CDE.

7 Dorothy Emmet, *Rules, Roles and Relations* (London: Macmillan, 1966), p. x.

8 Exceptions are Stephen Turner, 'Robert Merton and Dorothy Emmet: deflated functionalism and structuralism', *Philosophy of the Social Sciences*, 44:6 (2014), 817–36; and Peter West, 'Seeing life steadily: Dorothy Emmet's philosophy of perception and the crisis in metaphysics', *British Journal for the History of Philosophy* (2023), doi: 10.1080/09608788.2023.2256363.

9 Emmet, *Philosophers and Friends*, p. 1.

10 Kathleen Philip, *The Shy Village: Glimpses of the Story of West Hendred Village* (Wantage: K. Philip for the Hendreds Society, 1975).

11 Emmet, *Philosophers and Friends*, p. 1.

12 Ibid., p. 1.

13 See Clare Mac Cumhaill and Rachael Wiseman, *Metaphysical Animals: How Four Women Brought Philosophy Back to Life* (London: Chatto, 2023), pp. 18–19.

14 Ibid., p. 20.

15 Emmet, *Philosophers and Friends*, pp. 11–12.

16 Emmet, 'Memories of a Balliol stepdaughter', *Balliol College Annual Record* (1986), 47–50.

17 Emmet, *Philosophers and Friends*, p. 31.

18 Ibid., p. 79.

19 UML, VCA/7/227, 2/3, reference letter for DE by John W. Harvey, 6 March 1945.

20 Tadeus Szubka, 'An interview with Dorothy Emmet', *Cogito*, 8:2 (1994), 119. Contrast Wittgenstein's *Tractatus*: 'The world is the totality of facts', 1.1.

21 Emmet, *Philosophers and Friends*, p. 38.

22 Ibid.

23 Ibid., p. 34.

24 Lucy Cavendish College Archive, Cambridge (henceforth DME), LP17/3/2, letter from DE to Gertrude G. Emmet, 30/09/1938.

25 Emmet, *Philosophers and Friends*, p. 51.

26 DME, LP/17/2, *News Chronicle* report (probably December 1966).

27 DME, LP/17/5/2, 'Resolution of the Senate and Council of the University of Manchester recording their appreciation of the services rendered to the University by Emmet'.

28 Emmet dedicated her penultimate book, *The Role of the Unrealisable*, to Iris Murdoch, much to the latter's delight (DME, LP17/10/4 IM, undated letter from Iris Murdoch to DE).

29 Emmet, *Philosophers and Friends*, passim.

30 Ibid., p. 77.
31 VCA/7/227 2/3, letter from DE to vice-chancellor, 22 November 1937.
32 Emmet, *Philosophers and Friends*, p. 51.
33 Dorothy Emmet, *Function, Purpose and Powers* (London: St Martin's Press, 1958), p. 2.
34 Emmet, *Philosophers and Friends*, p. 76.
35 Ibid., p. 77.
36 Dorothy Emmet, *The Nature of Metaphysical Thinking* (London: Macmillan, 1945), p. v. For more on the crisis, see Mac Cumhaill and Wiseman, *Metaphysical Animals*, ch. 1.
37 A. J. Ayer, *Language, Truth and Logic* (1936) (London: Penguin, 1971), p. 27.
38 Emmet, *Nature of Metaphysical Thinking*, ch. 2. For discussion, see West, 'Seeing life steadily'.
39 Mary Midgley, *Owl of Minerva: A Memoir* (London: Routledge, 2005), p. 155. Midgley was born fifteen years after Emmet, but the two were remarkably similar in background and philosophical temperament, and met at least once: see Mac Cumhaill and Wiseman, *Metaphysical Animals*, p. 176, and Midgley, *Owl of Minerva*, p. 173.
40 Emmet, *Nature of Metaphysical Thinking*, p. 26.
41 H. L. A. Prichard, quoted in Matthieu Marion, 'Oxford Realism: knowledge and perception I', *British Journal of the History of Philosophy*, 8:2 (2000), 308.
42 Emmet, *Nature of Metaphysical Thinking*, p. 20 and *passim*. See A. D. Lindsay, 'What does the mind construct?', *Proceedings of the Aristotelian Society*, n.s. 25 (1924–25), 1–18.
43 Dorothy Emmet, *The Role of the Unrealisable: A Study in Regulative Ideals* (London: Macmillan, 1993), p. 64. I have had to leave unexplored here the connection between Emmet's objections to Ross and Prichard's Realism, and her objections to their moral intuitionism.
44 Emmet, *Nature of Metaphysical Thinking*, p. 26.
45 Szubka, 'An interview with Dorothy Emmet', 115.
46 Emmet, *Nature of Metaphysical Thinking*, p. v. It goes without saying that what follows is unforgivably compressed and simplified, and completely lacking in the subtlety and richness of Emmet's own discussion.
47 Emmet, *Nature of Metaphysical Thinking*, p. 26.
48 Ibid., p. 63.
49 Ibid., p. 26.
50 Ibid., p. 77. Donald MacKinnon was doing something similar at the same time, though within a more traditionally Thomist tradition. See his 'What is a metaphysical statement', *Proceedings of the Aristotelian Society*, n.s. 41 (1940–41), 1–26. As far as I know there is no comparative discussion of MacKinnon and Emmet.
51 Dorothy Emmet, 'The use of analogy in metaphysics', *Proceedings of the Aristotelian Society*, n.s. 41 (1940–41), 30.
52 Szubka, 'An interview with Dorothy Emmet', 116.
53 Emmet, 'The use of analogy in metaphysics', 39.

54 Emmet compares the metaphysician's 'total judgment' with the assertion that someone is 'a good man', as compared to the partial judgements that he 'looks after his old parents, gives away some of his eggs, spends his spare time on the APR' (Emmet, 'The use of analogy in metaphysics', 39).

55 Emmet, *Philosophers and Friends*, p. 78.

56 Emmet, 'The use of analogy in metaphysics', 38.

57 VCA/7/227 3/3, letter from Polanyi to vice-chancellor, 2 March 1945. See Emmet, *Philosophers and Friends*, pp. 74–84; for more on Polyani and Emmet's friendship, see *Philosophers and Friends*, pp. 74–84; and Jonathan Swinton, *Alan Turing's Manchester* (Manchester: Infang, 2019), pp. 87–92.

58 VCA/7/227 3/3, letter from John W. Harvey to vice-chancellor, 6 March 1945.

59 VCA/7/227 3/3, letter from G. C. Field to vice-chancellor, 6 March 1945.

60 VCA/7/227 3/3, letter from W. D. Ross to vice-chancellor, 8 June 1945.

61 Emmet, *Philosophers and Friends*, p. 88.

62 Dorothy Emmet, 'On the idea of importance', *Philosophy*, 21:80 (1946), 234–44.

63 See, for example, the opening of Austin's *Sense and Sensibilia* (Oxford: Oxford University Press, 1962), aimed at Ayer.

64 Mary Warnock, *A Memoir: People and Places* (London: Duckworth, 2000).

65 Emmet, *Philosophers and Friends*, p. 88. Here she and the legal philosopher H. L. A. Hart are fellow travellers (see, for example, Hart's *Concept of Law* [London: Clarendon Press, 1961]).

66 Emmet, *Philosophers and Friends*, p. 90.

67 Ibid., p. 90. It would be fascinating to look at Emmet's Whiteheadian influence in Gluckman's, MacKenzie's and Devons's books from this period, and on the work of the Manchester School more widely.

68 Swinton, *Alan Turing's Manchester*, pp. 66–95.

69 W. J. M. MacKenzie, *Politics and Social Science* (London: Penguin, 1967), preface.

70 Emmet, *Philosophers and Friends*, p. 90.

71 MacKenzie, *Politics and Social Science*, preface.

72 An anonymous reviewer in the *Economist*: 'Professor Emmet once seemed to have, to use her own word, a "vocation" for metaphysics. She did her own training with an admirable book on "The Nature of Metaphysical Thinking". But somehow she faltered when confronted with the task of doing metaphysics herself' ('Philosophy among the tribes', *The Economist*, 25 May 1958).

73 Emmet, *Philosophers and Friends*, pp. 96–7.

74 Again, there are many crucial aspects to Emmet's critique on which I will not touch. Of particular significance to social scientists is her claim that function should be treated as a heuristic, rather than explanatory, device. See Turner, 'Robert Merton and Dorothy Emmet'.

75 Bronislaw Malinowski, quoted by Emmet in her 'Functionalism in sociology', in Paul Edwards (ed.), *Encyclopedia of Philosophy* (London: Collier-Macmillan, 1967), vol. 3, p. 256.

76 For example, Emmet, *Function, Purpose and Powers*, pp. 45–7.

77 Emmet, *Function, Purpose and Powers*, pp. 260–3; Emmet, *Rules, Roles and Relations*, p. 149.

78 Emmet, *Philosophers and Friends*, p. 122.

79 Sartre, *Being and Nothingness*, quoted in Emmet, *Rules, Roles and Relations*, p. 153.

80 The moral philosopher Iris Murdoch, who was a great admirer of Emmet, observed this same tendency in the moral philosophy that dominated in British philosophy after the Second World War – for example, in R. M. Hare. See Iris Murdoch, 'The novelist as metaphysician' (1950) and 'The existentialist hero' (1950), in Iris Murdoch, *Existentialists and Mystics: Writings on Philosophy and Literature*, ed. Peter Conradi (London: Allen Lane, 1997), pp. 101–7 and 108–15.

81 Emmet, *Rules, Roles and Relations*, p. 153.

82 As Max Gluckman's travels yielded *Custom and Conflict in Africa* (Oxford: Blackwell, 1970).

83 Emmet, *Function, Purpose and Powers*, p. 256.

84 Ibid., p. 248.

14

Gilbert Gadoffre: institutionalising cultural reproduction

Derek Robbins

Mary Douglas (1921–2007) published *How Institutions Think* relatively late in her career as a social anthropologist. The book addressed explicitly an issue that she had tackled implicitly in all her work. She argued that 'a theory of institutions that will amend the current unsociological view of human cognition is needed, and a cognitive theory to supplement the weaknesses of institutional analysis is needed as well'.[1] She added:

> The whole approach to individual cognition can only benefit from recognizing the individual person's involvement with institution-building from the very start of the cognitive enterprise. Even the simple acts of classifying and remembering are institutionalized.[2]

Douglas saw institutions as objectified manifestations of individual volitions which themselves are conditioned by prior institutional forms. There is an affinity between the position adopted by Douglas and that developed in the same period in France by Pierre Bourdieu. His *Homo Academicus* explored the process of constitution of the 'academic field', and his *La noblesse d'état* analysed the institutionalisation of the *grandes écoles*, reflecting on the extent to which they were the products of state authority or of the self-determination of staff.[3] He drew on research he had conducted much earlier with Jean-Claude Passeron on student experience in the French higher education system. Their book, *La reproduction*, argued that the curriculum was a construction and imposition of those possessing arbitrary social power – that the transmission of a dominant culture

279

by the socially dominant necessarily led to the system excluding, or failing, those disadvantaged students who possessed valid, indigenous cultures.[4]

Gilbert Gadoffre (1911–95) was appointed to a chair in French at the University of Manchester, commencing in 1967, at a time when the British government was promoting expansion in higher education following the Robbins report of 1963. He had been an assistant lecturer at the University from 1938 to 1940 and then a special lecturer from 1954 to 1963. Throughout his career, Gadoffre was involved in developing new institutional forms within which to transmit the humanist values which he found especially in French Renaissance literature. The purpose of this chapter is to explore the dialectic between institutional and individual interest in his career and to consider the way in which his cross-cultural transfer of intrinsically French ideals impinged upon the English university institution with which he was associated in various ways for forty years (1938–78).

Before Manchester

Gilbert Gadoffre was born in Paris in June 1911. His father had served in Indo-China and spoke Chinese fluently. He was severely injured in 1914 in the Belgian campaign and died in a military hospital in Nice in June 1916. In 1917 Gilbert's mother took the family back to Paris where he started school. He developed pleurisy, with the result that his mother took the family to Switzerland to aid his convalescence. They were accompanied by a governess who played a large part in Gilbert's early development. She was from Lorraine, 'very anglophile, anti-German, and very musical'.[5] On returning to Paris, the family moved to Saint-Cloud, where Gilbert attended a local lycée before transferring to the lycée Hoche at Versailles. The family moved to Versailles, where there was a 'microsociety' of people who had made their careers in China.

Gadoffre's mother offered no resistance when he decided to break with the family tradition of military careers so as to study for a literature degree. When he was 12 and, again, when he was 19 he spent time in England as a paying guest. He was 'passionate about England'.[6] His father had known the rector of Lincoln College, Oxford, and Gadoffre studied there for one term, during which he made comparisons between the education there and at the Sorbonne that were all to the advantage of Oxford. Unlike at the Sorbonne, he thought that at Oxford 'students and teachers inhabited the same universe', taking 'tea together in groups'.[7] On returning from Oxford, he began two years of preparation for his *agrégation*, but his studies and his military service were both interrupted by acute pleurisy. He was able to get back to study only in 1936.

Gadoffre spent a year in the early 1930s working at a preventorium at Mégève in the Haute-Savoie, which had been established by a 'sort of St Vincent-de-Paul of modern times'.[8] Between 1936 and 1937 he volunteered there for three months, and he also spent the whole of 1937–38 there. He made friends, with the result that he was invited to spend most January months skiing in the years before and after the war. He wrote music criticism for *La Vie intellectuelle* and other journals between 1932 and 1937 under a nom de plume.[9] In this context he became associated with several significant Dominicans.[10]

Gadoffre had started research for a thesis on 'the image of China in French literature from 1884 to 1914'. He discussed Chinese issues as well as musical ones with his Dominican friends. Gadoffre recalls that the Dominicans 'represented for me a harmonious community which made visible and tangible that dream of community which I had always had'.[11]

In summary, it is possible to highlight several key factors in Gadoffre's early life. His fragile health often prevented or interrupted his formal education. He was a largely unschooled scholar. Perhaps the loss of his father at an early age also caused him to find solace in communities, either in the Franco-Sino society in Versailles, or in the collegiality of an Oxford college, or in association with Dominican priests. His interest in Chinese life and thought was in part indicative of respect for the father whom he had barely known, but, like his Dominican mentors, he was also attracted to Taoism and meditation. The military family background, the private schooling, the anglophilia and the Dominican connections all indicate a background that placed him outside the mainstream secular republican educational establishment of the French Third Republic.

Manchester I

The rector of Lincoln College, Oxford, recommended Gadoffre to Eugène Vinaver, head of the Department of French at the University of Manchester, when he was seeking a French member of staff. Gadoffre was offered a post which he took up in September 1938. He was given the task of teaching courses on French thinkers of the sixteenth and seventeenth centuries – Montaigne, Pascal and Descartes. Vinaver was directing a collection of texts entitled *Les ouvrages de l'esprit* [Works of the mind]. Ten volumes were published in French by Manchester University Press between 1939 and 1978. Vinaver asked Gadoffre to produce an edition of Descartes's *Discours de la méthode*. Gadoffre had completed a preface for this, but not his critical notes, when he was called up for military service in 1940. Believing that Gadoffre had been killed in June 1940, the University published the preface in memoriam in 1941. The complete edition was subsequently published in 1961.

Gadoffre's introduction was special in that he refused to respond as a philosopher to the text of Descartes. Instead, he outlined the context of the work's production, concentrating on Descartes's education at the Jesuit Collège de la Flèche and his interest in contemporary science. Gadoffre situated Descartes in the context of the 'generalised crisis of confidence' experienced in France following the scepticism of the end of the sixteenth century, which was the legacy of the wars of religion of that century.[12] He attributed the genesis of Cartesian thought to the pedagogical practice of the Collège, which was not dogmatic. It emphasised 'dispute' – not arid disputation in class but, rather, 'everyone disputed in class, while walking, while relaxing, in the lecture theatre, in any place, at any time, everywhere'.[13] We can sense Gadoffre's feeling of affinity with Descartes when he says of him that the Collège de la Flèche, 'to which the education of this young man, motherless and of delicate health, had been entrusted, was considered to be the best in France'.[14]

LES OUVRAGES DE L'ESPRIT
Collection de textes dirigée par Eugène Vinaver

DESCARTES

DISCOURS

DE LA MÉTHODE

avec introduction et remarques de
GILBERT GADOFFRE

EDITIONS
DE L'UNIVERSITE DE MANCHESTER
1 9 4 1

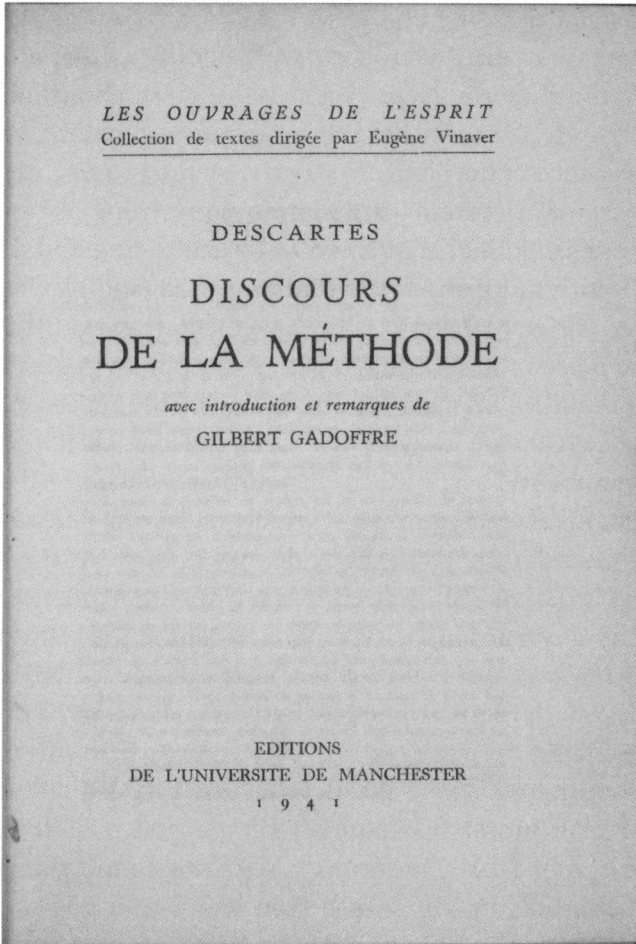

14.1 Gilbert Gadoffre's edition of Descartes was published by the University in 1941 as a memorial to him, in the belief that he had been killed in June 1940. The edition was awarded the Grand Prix Barthou by the Académie Française in 1941. Courtesy of the University of Manchester.

Intermission I

Gadoffre had, in fact, been wounded in action and made a prisoner of war. His left arm was paralysed for a year, which rendered him useless both to the Germans, who released him, and to the Free

French. Instead of being able to return to England as he wanted, from January 1941 until October 1942 Gadoffre taught at the lycée Condorcet and then the lycée Hoche, while living with his mother in Versailles. He led a double life, routinely smuggling explosives for the Resistance in his briefcase when travelling to give his lectures. He was part of a system of resisters performing various tasks, including assisting English airmen to return to England. He was in receipt of information and knew that an Allied landing was planned for October or November 1942. In the expectation that this landing would take place in the south of France, Gadoffre was sent to Uriage to help to mobilise support for the Allies in the region.

L'école des cadres à Uriage (the Uriage 'leadership' school) had been started in mid-1940 by the captain of a motorised squadron – Pierre Dunoyer de Segonzac – which had been defeated by the advancing German troops in the Ardennes in May. At the armistice he quit the army, and presented himself to the Ministry of Youth of the Vichy government with a plan to renew the nation's youth in preparation for a revival of the country. He recruited a team of like-minded staff, requisitioned a chateau near Vichy, and began to run 'stages' (training courses) for young men. The intention was that the training would be intellectual and physical. The purpose was to effect a moral rearmament of the youth of the country. Before the end of 1940, Dunoyer de Segonzac found the proximity to the government in Vichy to be too oppressive and controlling. The team moved to Uriage, near Grenoble, in November 1940, where it established a self-sufficient community practising manual skilled labour as well as participating in intellectual discussion which was stimulated in part by guest speakers. Gadoffre stayed at Uriage for ten days in August 1942, long enough to satisfy himself that the staff of the *école* 'were all anti-German, anti-Nazi, and allergic to every kind of collaborationism'.[15] Gadoffre gave an initiatory guest lecture in late July 1942, entitled 'The Crisis of Man and Humanism', and accepted an invitation to join the team from October.

Already the activities of the *école* were displeasing the Vichy government, and its closure was announced as from 1 January 1943.

The original team dispersed, but a rump of about twelve people established a new base in a chateau in Murinais, near to Saint-Marcellin, on the edge of the Vercors massif. The location became known among the team as 'the Thébaïde', connoting the priestly order in ancient Egypt. Under the continuing general direction of Dunoyer de Segonzac, Gadoffre led the team at Murinais. The original lifestyle was maintained, but the new team became involved in 'flying visits' to support the Resistance groups in the Vercors while also focusing on the task of giving coherent intellectual form to the ideas that had been discussed at Uriage. Gadoffre was almost alone in the chateau when, on 13 December 1943, German troops and the Milice burned it down. The drafts of Gadoffre's thesis on Paul Claudel and of the 'summa' of the ideas of Uriage under his editorship were both destroyed in the conflagration. Gadoffre escaped in dramatic fashion, as he described in the novel he wrote about his wartime experiences, *Les Ordalies*, published in 1955.[16]

After his escape from Murinais, Gadoffre managed to get to Paris, where he stayed in a mansard room for three months, working, in clandestine collaboration with other members of the team, on the restoration of the text of the summa which was to be published in 1945 as *Vers le style du XXe siècle* [Towards a style for the twentieth century].[17] Dunoyer de Segonzac then arranged for Gadoffre and some others to be lodged in a chateau near the maquis at Grammont to finalise the text before they joined him in the Tarn. Gadoffre was made a captain in the French Forces of the Interior. He was in the Black Forest when the war ended, reaching Constance in August 1945.

Vers le style, edited by Gadoffre and published in 1945, was a complex integration of the differing pedagogical and political attitudes of the team of authors. They loosely shared a conviction that the necessary revolution in society could not be effected by politicians and they were strongly internationalist (while tacitly assuming that the reformation of French society would necessarily supply the model for world transformation). Other members of the team believed in the encouragement of popular culture, supporting working-class, popular or indigenous cultures either by developing pioneering teaching methods or by making the case for autodidacticism.[18] Gadoffre automatically assumed that cultural exchange was a matter of 'high' culture dialogue. In his contributions to *Vers le style*, Gadoffre outlined his vision for the introduction of 'maisons de la culture' throughout France. These were to be developed differently in different kinds of socio-economic context – some in 'cities of art' such as Avignon and some in 'industrial cities' and 'devastated regions' – but they were conceived by him as institutions for the top-down dissemination of culture to the culturally deprived.

From July 1945 to November 1946 Gadoffre was director of information in the French zone of Austria. A society of Friends of France was established, and Gadoffre played a part in trying to mitigate the effects of the French occupation and to ensure that Austro-French relations were restored, but he soon sensed that 'the reconstruction of France was taking place without me'.[19] It must therefore have been a relief when Beuve-Méry, a former member of the Uriage team and then editor of *Le Monde*, put him in touch with the owner of the old Cistercian abbey at Royaumont to the north of Paris, which enabled Gadoffre to set about realising some of the projects outlined in *Vers le style*. At Royaumont, in 1947, he realised the plan for an international institute for French civilisation.[20] The buildings and grounds at Royaumont met the conditions which, in *Vers le style*, he envisaged would be met at the chateau of Blois. He envisaged that Blois

> should become a truly studenty city, swarming with the most diverse activities and capable of being self-sufficient. Conceived in this way university life should not only be an initiation into a culture and a profession but also a context for civic and social education, a total preparation for the profession of being a man.[21]

Gadoffre stayed at Royaumont from May to Christmas each year until 1954. He organised courses for foreign students, initiating them into French culture and French university life, colloquia to which he invited prestigious guests of all disciplines and from many nations, and musical and theatrical events. Constantly he tried to juxtapose the humanism of sixteenth-century Europe with that which he thought to be necessary in the postwar world.

In 1953 Gadoffre married and he also decided that he would like to take up the university career that had been interrupted by his mobilisation in 1940. In 1954 he was asked by Vinaver to return to Manchester to teach the 'history of French thought'.

Manchester II

The Manchester University to which Gadoffre returned had recently very consciously reaffirmed its allegiance to the values of its founder, John Owens. An account of the history of the institution (1851–1951) written by its Professor of English Literature H. B. Charlton argued that Owens believed that social progress was to be secured through education. The central thread of Charlton's centenary volume was 'to follow a new educational concept fulfilling itself in response to emergent circumstance, to see the evolving idea of a new form of university, a university with a new conception of its place and its duty to society'.[22] This view was endorsed by the vice-chancellor, Sir John Stopford, in his preface. The text published letters of congratulation sent to Manchester from the University of Paris on the occasion of the Jubilee in 1901, in which the vice-rector, president of the University Council and the secretary to the Council, Ernest Lavisse, commented on Manchester University's good fortune in having been born 'in one of the capitals of modern industry', and claimed that the world was indebted to Manchester's pioneering demonstration that 'speculative and aesthetic thought' should not be divorced from scientific and technological advance.[23] Manchester was indisputably an advanced industrial city and the University was committed to reflecting this ethos.

Gadoffre recalled that he and his wife at first installed themselves in a hotel in Victoria Park, and that when they returned to it in the evenings 'along pathways full of muddy holes and when the fog made it difficult for us to see our way', he had regrets that he had 'brought his Corsican wife to such a place'.[24] Later they moved out to Cheshire, where his wife was happier to settle.[25]

Gadoffre's responsibility was to teach the 'history of French thought'. He endorsed Vinaver's contention that no training in French is possible without 'an initiation into this central structural support [*armature*] of French literature' and he maintained that this was a distinctive feature of the Manchester French Department.[26] He strongly believed in the civilising effect of immersion in the work of sixteenth- and seventeenth-century French writers and philosophers as autonomous cultural practices.

It must be remembered that Gadoffre in these years contrived to spend six months in each year in England and share the third term between Germany and France. In France he carried on organising colloquia each year for the Collegial Institute which he had founded at Royaumont in 1949. Until 1959 he ran these colloquia in different locations in France, while Versailles was the 'centre of gravity'. One colloquium of 1955 was particularly significant. Entitled 'L'après-guerre est-il terminé?' [Is the 'post-war' period over?], it hosted a dialogue between Nathalie Sarraute and Alain Robbe-Grillet which provoked the latter's announcement of the advent of the 'new novel'. Consistent with his orientation at Uriage, Gadoffre was sympathetic to the view that a new, postwar society required a new novel for the future, one that rejected the Balzacian model.

In *Gilbert Gadoffre, un humaniste révolutionnaire*, Gadoffre's wife asked him whether he had felt the need to become politically engaged. The question is a prelude to his reflections on the career of Pierre Mendès-France, whom he had supported when he was Radical Party prime minister in 1954–55. Gadoffre was 'traumatised' when Mendès-France failed to defeat de Gaulle in 1958. From that moment he 'never invested in a politician'.[27] It was in relation to the reforms proposed by Mendès-France in 1958 that Gadoffre met André Malraux and encouraged him to implement the 'maisons de la culture' project.

While organising his Collegial Institute event of 1959 in Touraine, Gadoffre received the support of the sub-prefect, such that he decided to establish a new permanent centre at Loches-en-Touraine. This lasted until 1993. Unlike the situation at Royaumont, that at Loches made possible an integration of the cultural centre with 'a true small town, with a true society'.[28] It was not 'a lost place where intellectuals gathered'.[29] He and his wife lived there from mid-June until mid-October each year and organised on average two colloquia each year, their guests arriving from all over the world and staying on to be entertained for 'after-colloquia'. Not all the guests could be accommodated at the centre, and Gadoffre's memory is of sessions conducted in a community that also incorporated the actual community of the town. As a consequence of the history and architecture of the town, Gadoffre felt that what was important was that the location gave everyone 'the impression of living through superimposed epochs, moving from the twelfth century to the Renaissance, bathing in a quintessence of history and civilisation'.[30]

Intermission II

While Loches was developing, Gadoffre received an invitation to teach at the University of California, at Berkeley. He stayed from 1963 until 1966. As early as 1947/48, Gadoffre had published articles in *Le Monde* containing reflections on French, American and English universities.[31] He was sure that institutional change was necessary to achieve an adequate response to the challenges of mass democracy. He thought that it was his book on Ronsard that had stimulated American interest in appointing him, but he found that he was required to offer a course on nineteenth-century French literature – a period with which he was not intrinsically in sympathy.[32] He chose to use this to reflect on the social context of literary production in the period, focusing on the ways in which Lamartine's public readings of his *Méditations* appealed to an audience lacking prior education. He also took the opportunity to introduce students to the development of the Bateau-Lavoir group in Montmartre at the beginning of the twentieth century, which had assembled artists such as Picasso and poets such as Apollinaire. Both of these emphases involved historical analysis of the relationship between 'pop culture' and 'high culture'. This was Gadoffre's indirect engagement with the emergent youth culture in American universities in the 1960s. Gadoffre was attracted by the liberalism of American universities and by the vitality of the youth movement, which seemed to him to realise in some degree his earlier desire to animate a 'new style' for the twentieth century. He preferred what he saw as the cultural dimension of student protest in the United States to the later sociopolitical orientation of the 'May events' of 1968 in Paris. In retrospect, Gadoffre reflected that it had not been difficult to suggest to his American students that the great advantage of 'cultural foyers' such as Lamartine's chateau and the Bateau-Lavoir was that they made it possible to 'distinguish culture from the civilization that supported it'.[33]

Manchester III

In May 1965 the vice-chancellor at Manchester reported that Professor Vinaver would retire at the end of September 1966. The Senate established a committee to consider future arrangements in French Language and Literature. This committee made some significant resolutions. It decided that Vinaver's chair, which had been of French Language and Literature, should be renamed the chair of Classical French Literature and be given to the existing chair of Modern French Literature, with the result that it argued that the new appointment should go to a specialist in the literature of the nineteenth and twentieth centuries. It resolved, secondly, that the vice-chancellor should write to designated specialists in the field to seek their recommendations for possible candidates. The chosen specialists were professors of French in other UK universities and also the rector of the Académie at Orleans and the professor at the University of Strasbourg, who was also the French cultural counsellor in the UK. Notwithstanding the fact that one of the referees doubted Gadoffre's fluency in spoken English and also that it was generally admitted that his specialism was not primarily in relation to French literature of the nineteenth and twentieth centuries, he was offered the post without any competitive interview, returning from the USA to commence in 1967.[34]

```
CURRICULUM VITAE

        Assistant-lecturer in French literature in Manchester Uni-
        versity (1938-1940)

        WAR : Mobilised in the French army (April 1940); wounded
        (June 1940);Military cross; Member of an underground move-
        ment from March 1941 ; maquis instructor (1943-44);Condemned
        to death "in absentia" by German military authorities.Manus-
        cripts and material of pre-war research seized by the Gestapo
        in December 1943 ; Captain in the first French Army (1944-
        1945).Médaille de la résistance.

        Director of the cultural information office in the French
        zone of Austria (1945-1947)

        Director of the Institut Collégial européen (International
        round table conferences set up in Royaumont from 1947 to
        1954, and then in the château of Loches-en-Touraine and
        Versailles )

        from 1954 onwards : Sp.Lecturer in the history of French
        thought (XVIe and XVIIe centuries) at Manchester University

        from 1964  Professor in French litterature at the Universi-
        ty of California (Berkeley)
```

14.2 An extract from Gadoffre's CV from his application for the chair of Modern French Literature at Manchester, 1965. It records that his pre-war research materials were seized by the Gestapo in December 1943. Courtesy of the University of Manchester.

Gadoffre remained at Manchester until his retirement in 1978. *Gilbert Gadoffre, un humaniste révolutionnaire* gives no indication of the kinds of courses he taught during those eleven years.[35] He had begun a thesis on 'Claudel et l'univers chinois' [Claudel and the Chinese universe] in the late 1930s. A draft was destroyed in the fire at Murinais in 1944, but Gadoffre revived it to secure both an English and a French doctorate before the book was published in 1968.[36] This could be taken as a legitimation of Gadoffre's qualification to teach modern French literature, in that Claudel (1868–1955) was a major modern French poet and dramatist, but Gadoffre's book explored the relationship between Claudel's literary activity and his professional life as a diplomat in China from 1895 until 1909 (including the period of the Boxer Rebellion of 1900–01). Early Claudel did not see the arrival of Western nations as beneficial to Chinese culture, and Gadoffre's attention to this suggests an affinity with Claudel's view, a correspondence with the decline of traditional French society.[37]

Gadoffre's main academic concern throughout this period of his life at Manchester seems to have been to represent the work of the 'Pléiade' – the poets (notably Ronsard and Du Bellay) who, in the sixteenth century, had issued a manifesto undertaking to reassert the values of classical learning and literature in the vernacular.[38] His book on Ronsard (1524–85) had highlighted that he had been born in an age when the values of humanism and the place of the humanist in society were in question, and that the recovery was the consequence of the domination of the colleges over the faculties in the medieval university.[39] His book on Du Bellay (1522–60), published in 1978, followed several articles on the subject earlier in the 1970s.[40] Here Gadoffre concentrated on the tensions between a re-emerging humanism and a conservative Catholic establishment. Gadoffre always sought to emphasise the need for society to generate a new humanism for changing circumstances, but sought equally to recommend the *content* of French Renaissance culture more than its *form*.

Throughout his time at Manchester, Gadoffre's activities in France continued, and, after retirement, he was involved in setting up an ongoing interdisciplinary seminar at the Collège de France, which he co-directed from 1979 until 1988. He saw this as a realisation of the project in *Vers le style* for the establishment of an Institute of Synthesis.[41]

Gadoffre's influence came mainly through his endeavours in a kind of intellectual entrepreneurialism as well as through his university teaching. Additionally, however, he produced a series of books as well as overseeing the production of many edited proceedings of events and colloquia. In the foreword to *Un humaniste révolutionnaire*, Alice Gadoffre-Staath tells how, shortly before his death in March 1995, he was finishing a book which he had been preparing for twenty years. It was published posthumously as *La Révolution culturelle dans la France des humanistes* [The cultural revolution in the France of the humanists].[42] Perhaps as an ultimate, self-regarding reflection, Gadoffre emphasised the roles of Guillaume Budé, administrator, and François I, monarch, in the advancement of sixteenth-century French humanism.

Conclusion

Based on the slim documentary evidence to hand, it is difficult to avoid the sense that Gadoffre's mind was anything more than tangentially engaged with his situation in Manchester. Although he proclaimed himself to be passionately anglophile, almost all his books and articles were written and published in French. Until he disowned 'politicians' in the late 1950s, he was actively involved in French postwar politics and, thereafter, until his death, he was concerned with French cultural projects as potential instruments for political change. In England, he saw himself as a cultural attaché – as he had been in Austria in 1946–47 – rather more than as a university professor.[43] His commitment was always to the fostering of cultural exchange within socially autonomous, holistic environments. This was realised in institutional contexts of his own making rather than in pre-established formal educational situations – perhaps as a consequence of the historical contingency of his upbringing. It is clear, nevertheless, that Manchester students came under his influence for the best part of forty years. One student from 1959–60 recalls that Gadoffre gave 'a series of lectures in French entitled "Introduction à la pensée française", which were in fact nothing less than a summary history of Western Philosophy from the Presocratics to Montaigne', and that he was 'fascinated by the clarity of the connections made and the likenesses and differences between different lines of thought'.[44]

The committee that implemented the will of John Owens of 1845 in founding the originating college of Manchester University specified five primary teaching subjects to be established. These comprised 'Classical Languages and Literatures; mathematics; natural philosophy or physical science; Mental and Moral Philosophy; English Language and Literature (with main emphasis on the discipline of general grammar)'.[45] It stated that it was prepared to add 'as useful secondaries, Natural History and some of the modern languages'.[46] The first chair of French Language and Literature was established in 1895, an indication of the cultural arbitrariness of curriculum development.[47] In our age of international student mobility, Gadoffre's career forces us to ask whose 'culture' should be transmitted to whom in our universities, and why.

* * *

This account has been guided by Gadoffre's self-presentation edited by his widow, Alice Gadoffre-Staath, and published in 2002 after his death as *Gilbert Gadoffre, un humaniste révolutionnaire* [Gilbert Gadoffre, a revolutionary humanist]. This was based on transcripts of his late conversations and also contains excerpts from some of his publications as well as some appreciations of his career. I have also attempted to incorporate into my account some interpretation of Gadoffre's published works, understood in relation to the contexts of their production. Traces of Gadoffre in the University's archives are few, but I am grateful to Stuart Jones for tracking down some direct archival records relating to Gadoffre's time at Manchester.

Notes

1 Mary Douglas, *How Institutions Think* (London: Routledge and Kegan Paul, 1987), p. ix.
2 Ibid., p. 67.
3 Pierre Bourdieu, *Homo Academicus* (Paris: Minuit, 1984) and, in English, *Homo Academicus*, trans. Peter Collier (Cambridge: Polity, 1988); Pierre Bourdieu, *La noblesse d'état. Grandes écoles et esprit de corps* (Paris: Minuit, 1989) and, in English, *The State Nobility: Elite Schools in the Field of Power*, trans. Lauretta C. Clough (Cambridge: Polity, 1996).
4 Pierre Bourdieu and J.-C. Passeron, *La reproduction. Éléments pour une théorie du système d'enseignement* (Paris: Minuit, 1970) and, in English, *Reproduction in Education, Society and Culture*, trans. Richard Nice (London: Sage, 1977).
5 Alice Gadoffre-Staath (ed.), *Gilbert Gadoffre, un humaniste révolutionnaire* (Paris: Éditions Créaphis, 2002), p. 11.
6 Ibid., p. 15.
7 Ibid., p. 15.
8 Ibid., p. 17.
9 *La Vie intellectuelle* was a Catholic review created in 1928 by a Dominican priest at the request of Pope Pius XI and with the support of Jacques Maritain.
10 For further details, see Derek Robbins, *Towards a New Humanity: The Uriage Manifesto, 1945* (Oxford: Peter Lang, 2021).
11 Gadoffre-Staath (ed.), *Gadoffre*, p. 21.

12 Gilbert Gadoffre (ed.), *Descartes. Discours de la méthode* (1941) (Manchester: Manchester University Press, 1961), p. viii.

13 Ibid., p. xiv.

14 Ibid., p. xiii.

15 Gadoffre-Staath (ed.), *Gadoffre*, p. 26.

16 Gilbert Gadoffre, *Les Ordalies* (Paris: Seuil, 1955).

17 Gilbert Gadoffre (ed.), *Vers le style du XXe siècle* (Paris: Seuil, 1945). For the English translation, see Robbins, *Towards a New Humanity*.

18 I am thinking particularly of Joffre Dumazedier and Bénigno Cacérès, who together subsequently founded the *Peuple et Culture* movement.

19 Gadoffre-Staath (ed.), *Gadoffre*, p. 59.

20 Ibid., p. 185.

21 Robbins, *Towards a New Humanity*, pp. 215–16; Gadoffre (ed.), *Vers le style*, p. 191.

22 H. B. Charlton, *Portrait of a University, 1851–1951* (Manchester: Manchester University Press, 1951), p. 4.

23 Ibid., pp. 77–8. Ernest Lavisse (1842–1922) was a positivist historian and prominent advocate for educational reform in the Third Republic who was to be appointed director of the École normale supérieure in 1903.

24 Gadoffre-Staath (ed.), *Gadoffre*, p. 90.

25 Ibid., p. 90.

26 Ibid., p. 88.

27 Ibid., p. 101.

28 Ibid., p. 114.

29 Ibid., p. 114.

30 Ibid., p. 118.

31 These are reproduced in ibid., pp. 198–209.

32 'A century which does not especially attract me in respect of French literature': Gadoffre-Staath (ed.), *Gadoffre*, p. 130. The Ronsard book was Gilbert Gadoffre, *Ronsard* (Paris: Seuil, 1960).

33 Ibid., p. 131.

34 For the information in this paragraph, see the minutes of the committee meeting of 7 May 1965 and the attached correspondence in VCA-7-749.

35 Old examination papers in the University archives suggest that he was mainly involved in introducing students to the work of Montaigne, Descartes and Pascal.

36 Gilbert Gadoffre, *Claudel et l'Univers chinois* (Paris: Gallimard, 1968).

37 See the discussion in ibid., ch. VI, pp. 143–72.

38 See the account in Anne Denieul-Cormier, *The Renaissance in France 1488–1559* (London: Allen and Unwin, 1969), particularly ch. XI: 'Schools and scholars'.

39 Gadoffre, *Ronsard*, pp. 49, 27–8.

40 Gilbert Gadoffre, *Du Bellay et le sacré* (Paris: Gallimard, 1978).

41 This project is outlined in sub-section C of section IV of Part II, chapter 2 of Gadoffre (ed.), *Vers le style*; Robbins, *Towards a New Humanity*, pp. 211–16.

42 Gilbert Gadoffre, *La Révolution culturelle dans la France des humanistes* (Geneva: Droz, 1997).

43 An obituary of Gadoffre written for the *Guardian* of 30 March 1995 by Geoffrey Harris was entitled 'Cultural ambassador'.

44 From *Gilbert Gadoffre, a personal memoir*, in a letter from Tony James to Stuart Jones dated 1 February 2023.

45 Charlton, *Portrait of a University*, pp. 28–9.

46 Ibid., p. 29.

47 Ibid., Appendices.

PART III

The University in the post-industrial city

Stuart Jones

If, as the young Andy Spinoza thought, Manchester looked in 1979 like a city 'locked in a fatal post-industrial tailspin', the forty-five years since have seen a stunning reversal.[1] The city's population, which was consistently in decline over the second half of the twentieth century, started to grow again around the millennium: from 393,000 in 2001 to 503,000 in 2011 and around 552,000 in 2021. In the 2020s something not far short of 20 per cent of Manchester's population are students, which is about twice the density of the UK as a whole. Particularly striking was the rejuvenation of the city centre: it had a population of 500 in 1982, but had grown to 60,000 in 2023. The larger area defined by the city-region as its 'central core' – encompassing both Salford Quays and Trafford's 'civic quarter' – now exceeded a quarter of a million residents.[2]

Spinoza attributes the resurgence to the Manchester music scene of the 1980s: 'No Haçienda, no new Manchester'.[3] The city was certainly culturally resurgent by the end of the decade, but the economic revival probably owed much more to the city council's ambitious programme of regeneration following the IRA bomb in the city centre in July 1996. A landmark in the repositioning of Manchester as a thrusting contemporary city was the successful bid to host the Commonwealth Games, the athletes' village being located in the University's Fallowfield campus. Some major civic projects in fact pre-dated the bomb: the Bridgewater Hall, a major concert venue built with support from the European Regional

Development Fund, was largely completed before the bomb and was opened shortly afterwards. Equally, the Metrolink tram/light rail system pre-dated the bomb by several years. It was central to the creation of greater connectivity within Greater Manchester, and now has eight lines and ninety-nine stations. It bypasses the academic quarter, however: the University is a mile from the nearest station, and Fallowfield campus two miles.[4]

It was against this background of civic renewal that the merger took place in 2004 between the Victoria University of Manchester and UMIST. The latter had been, until 1994, formally a part (though a highly autonomous part) of the Victoria University, but what happened in 2004 was absolutely not a U-turn, but the self-conscious creation of a new university. As Luke Georghiou shows in Chapter 18, it was a deliberate choice that no institutional structures, regulations or policies should be regarded as sacrosanct. What emerged was a university with an ambitious strategy to reposition itself as a global leader. It was a vision that almost certainly could not have been generated in the declining city of a quarter of a century earlier, and it was strongly infused with a sense of the city's new-found buoyancy. Conversely, the formation of the new University, which entailed a major programme of campus redevelopment stretching over two decades, contributed powerfully to the renewal of the city and its reputation for self-conscious modernity. It was a new university for a renewed city, though both the city council and the University have periodically faced criticism for their alleged embrace of neoliberal politics.[5]

At an early stage the new University identified 'social responsibility' as a core goal, given equal weight alongside teaching and learning and research and discovery. It also identified 'civic engagement' as an activity that, in company with 'innovation' and 'global influence', should permeate all aspects of its work (see Epilogue, Figure e1.1). The contemporary University is influenced by a sense of its own heritage as the descendant of the institutions in which the very idea of the civic university was first shaped: indeed, its logo, highlighting the foundation date of 1824, stands in counterpoint to the brash novelty of its re-foundation in 2004. In fact the idea of the civic university has experienced a remarkable rediscovery in the last decade. We have seen not only the publication of the first comprehensive overview of the history of the English civic universities, but also a torrent of studies of the meaning of civic engagement for universities today.[6] The BPP Foundation established a Civic University Commission which produced two high-profile reports that helped inspire a series of 'Civic University Agreements' in different cities and city-regions. The first, appropriately enough, was in Greater Manchester (see Epilogue). This rediscovery of the civic university tradition has been rooted in a new sense of the role of universities as 'anchor institutions': institutions that are not only economically impactful, but also tied to a particular place, and hence central to projects of civic renewal – increasingly so, as local authorities have borne the brunt of successive waves of austerity. As the Civic University Commission noted in its final report, 'Universities have moved from being dependent on the cities in which they are situated, to being economic drivers of places in their own right.'[7]

In her Epilogue, Dame Nancy Rothwell makes a powerful case for the centrality of today's University – and of Greater Manchester HEIs considered together – to the economic vitality of the city-region. The relationship between the University and civic government is more akin to a business partnership, rooted in a shared awareness of the huge impact a large research university can have on the regional economy as well as on local communities. In the face of calls for universities to become 'truly civic', the challenge is to fuse the sense of the University's economic importance with a deeper sense that it 'belongs' to the city.

Notes

1 Andy Spinoza, *Manchester Unspun: Pop, Property and Power in the Original Modern City* (Manchester: Manchester University Press, 2023), p. 1.
2 Ibid., p. 2.
3 Ibid., p. 3.
4 In the case of the University, the distance is measured from the John Owens Building.

5 Jamie Peck and Kevin Ward (eds), *City of Revolution: Restructuring Manchester* (Manchester: Manchester University Press, 2002), especially Allan Cochrane, Jamie Peck and Adam Tickell, 'Olympic dreams: visions of partnership', pp. 120, 132; Isaac Rose, *The Rentier City: Manchester and the Making of the Neoliberal Metropolis* (London: Repeater, 2024).

6 William Whyte, *Redbrick: A Social and Architectural History of Britain's Civic Universities* (Oxford: Oxford University Press, 2015); John Goddard and Paul Vallance, *The University and the City* (Abingdon: Routledge, 2013); John Goddard, 'The civic university and the city', in Peter Meusburger, Michael Heffernan and Laura Suarsana (eds), *Geographies of the University* (Cham: Springer, 2018), pp. 355–74.

7 'Truly civic: strengthening the connection between universities and their places: the final report of the UPP Foundation Civic University Commission' (2019), https://upp-foundation.org/wp-content/uploads/2019/02/Civic-University-Commission-Final-Report.pdf (accessed 28 March 2024).

15

Making an impact: Brian Cox, Jodrell Bank and changing perceptions of science in the twenty-first century

Matthew Cobb

On 10 September 2008 the Large Hadron Collider (LHC) at CERN in Switzerland was switched on. This massive particle accelerator – the largest machine ever built – was designed to detect a fundamental subatomic particle, the Higgs boson. There was uninformed speculation that the LHC might create a black hole and destroy the world (it did not),[1] but the BBC was more upbeat and took the remarkable step of devoting virtually all of Radio 4's output to programmes about the event, under the title 'Big Bang Day'.[2]

The blanket coverage began with the *Today Programme*, co-presented from CERN, followed by a programme on the construction of the LHC. *Woman's Hour* then discussed how to involve more women in science, the relevance of which was demonstrated by the following programme, *Physics Rocks*, in which Brian Cox of the University of Manchester talked to 'celebrity physics enthusiasts', all of whom were men. After *The Archers* (which disappointingly did not have an LHC theme), the afternoon play was a *Doctor Who* spin-off in which 'something strange' took place as a giant particle accelerator was turned on; then the science of subatomic particles was explained by mathematician Simon Singh, while the evening arts programme *Front Row* explored the representation of physics in the arts (more men, including Tom Stoppard). Finally, the actor Ben Miller described how the LHC worked and what it was designed to detect, while the day's programming closed with a thirty-minute 'fast-moving comedy' set in the LHC.

Brian Cox was also on BBC 2's *Newsnight* programme that evening, where he clashed with Sir David King, the former chief scientific advisor to the UK government. King argued that the LHC would produce no future benefits to humanity; Cox – whose research was based at CERN – was incredulous and pointed out the many spin-offs that occur from 'blue skies' research (plate 6). As the *Observer* put it, Cox had 'leapt to ubiquity', and it seemed that 'television might have discovered its evangelist for science'.[3]

This was a key moment in a significant shift in media representations of science that took place in the first two decades of the twenty-first century. The immense scale of the LHC and the deep questions about the nature of the universe that it addressed made this an exciting moment, while the fact that much of the population finds particle physics simultaneously intriguing and highly abstruse meant that expert scientific interpreters and popularisers were needed to frame the research in simple terms. Before and after this event, reaching up and down the timeline, other factors contributed to the change in public attitudes to science, but the LHC was the pivot point.

This chapter traces how a combination of anniversaries, media attention, government policies and enthusiastic innovation by scientists – in particular by members of the University of Manchester – helped contribute to and shape a growth in interest in science, and in particular physics, around the world. This wave of enthusiasm saw the communication of science move away from a talking head Reithian approach towards a more playful yet serious blend of science, the arts and comedy. It was particularly personified by Brian Cox and his collaborators, but also by the staff at the University's radio telescope at Jodrell Bank, who regenerated and transformed the site into a centre for both science outreach and entertainment, culminating in the launching of the hugely successful bluedot science and music festival. In these decades, science came out of the University's laboratories and pushed its way on to the television screen and into the laser-filled night skies, to the pulsing sound of dance music.

The rise of Brian Cox

Cox might appear to have 'shot to ubiquity' in 2008, but in reality he had been building up his media profile for a decade, with the aim of making science part of popular culture. Cox famously gave up being in the rock band D:Ream to do his PhD in particle physics at Manchester, and together with his partner and future wife, television presenter Gia Milinovich, he had tried to interest broadcasters in a television programme that would mix science and popular culture, but the high-ups 'didn't get it'. However, by 2004 Cox had caught the attention of the BBC Radio Science Unit, and he was invited to present a three-part Radio 4 series, *In Einstein's Shadow*, to mark the centenary of Einstein's theory of special relativity, which was published in 1905; Cox also fronted five companion programmes of extracts from Einstein's letters, *Dear Professor Einstein*.[4]

Broadcast in January 2005, and produced by Alexandra Feachem, *In Einstein's Shadow* was the first programme that Cox had presented. The opening episode began with him interviewing his PhD supervisor and head of department, Robin Marshall, in the University of Manchester's Whitworth Hall, where in 1921 Einstein gave his first UK lecture (in German). The series also went to CERN, where the LHC was being built. Cox expressed his amazement as he clambered into a 'James Bond style lift' and descended 80 metres below the Swiss countryside, to come face to face with the vast space that would be filled by the ATLAS detector that he was working on. His awed, spontaneous and unaffected response to the 'industrial scale science' at the LHC soon became his media signature, grabbing the attention of audiences the world over.

At the beginning of 2007, as the LHC approached completion, the BBC made a TV science *Horizon* programme about the project, *The Six Billion Dollar Experiment*.[5] Cox opened the programme, striding into CERN and then riding an electric tricycle around the 27-kilometre circular tunnel that housed the accelerator in a massive blue pipe, grinning and telling the audience 'this is the most exciting place in all of science ... just look at it ... this blue

... you can see it's an exciting colour!' Being slightly overwhelmed and even tongue-tied revealed Cox as spontaneous and genuine, two magical televisual qualities.

Those early BBC programmes introduced Cox to two BBC producers who would work with him over the subsequent decades – Alexandra Feachem in radio and Andrew Cohen in television. Although to the viewer or listener the programmes were all about Cox, the producers and editors, as well as composers, camera operators and many others, played an essential part in creating his style and the illusion of effortlessness.

In the following months, Cox honed the informality that became his hallmark through a CERN podcast in which he chatted about the LHC and science in general to various personalities, including the Conservative MP Ed Vaizey (who, like Cox, had campaigned against cuts to physics funding), the Dean of Guildford Cathedral and the satirist Chris Morris.[6] As one of the commenters on the podcast blog put it: 'You are fast becoming the Carl Sagan of our time.'[7] At the same time, Cox was learning to explain complex ideas in a simple way by presenting a GCSE maths revision pro-gramme for BBC Bitesize, something which brought him to the attention of many grateful teenagers.

Cox next worked with Andrew Cohen on a *Horizon* programme entitled *What on Earth is Wrong with Gravity*, produced and directed by Paul Olding and broadcast in January 2008.[8] This marked a key step in the establishment of the Cox style – he travelled to various exotic locations in the USA (and to Isaac Newton's garden in Grantham), stood on hilltops and looked in awe at the sky, presenting the programme in a blend of the authoritative and the informal, sometimes corpsing and deconstructing the medium itself. Dem-onstrating physical principles in a café, talking to camera in an effortlessly natural way, and linking physics with the nature of the universe itself, the Cox style was being created.

Shortly before Big Bang Day, Cox fronted a relatively straightfor-ward BBC 4 documentary about the LHC, *The Big Bang Machine*, with Cohen as the executive producer. Such was public interest in the LHC that the programme was repeated three times and was seen by over a million viewers, over half of them under 34, with nearly 40 per cent of them women.[9] At the end of the year another *Horizon* programme, on the nature of time, *Do You Know What Time It Is?*, which was another collaboration with Cohen and Olding, cemented the Cox style established earlier in the year, from walking up a Mayan pyramid at sunrise to informally chatting to an off-screen interviewer. This programme won awards, and the combination of scientific clarity, a warm personality and a relaxed and modern *mise-en-scène* laid the foundations for Cox's subsequent fame. Another *Horizon* programme soon followed – *Can We Make a Star on Earth?*, about nuclear fusion, with Cohen as executive producer and long-term collaborator Gideon Bradshaw as producer.

While Cox clearly had the It Factor, these programmes did not grab a mass audience. This was initially even more true for his work for radio, but in early 2009 Feachem, with the support of Deborah Cohen (no relation), head of the BBC Radio Science Unit, tried to make a *'Top Gear* of science' – a radio show featuring a bunch of blokes sitting around talking enthusiastically. The unbroadcast pilot programme featured Cox, science journalist Adam Rutherford and medic Kevin Fong. It did not work. Despite their different accents, all three men basically sounded the same and above all they all said similar things.

However, the programme included a brief sketch written by comedian Robin Ince, who had an English literature degree but was interested in science, and in particular in physics. Ince's comical interactions with Cox, his irreverent tone and its contrast with the more sober demeanour of the scientists gave Feachem and Cox the idea for what became *The Infinite Monkey Cage*, a globally successful radio programme and podcast fusing science and comedy. In each episode Cox and Ince would invite guests – normally a pair of scientists and a comedian – to discuss a particular theme; the initial series, recorded in the studio, remained staid but the magic sparked when they performed in front of an audience at the Cheltenham Science Festival in June 2010. After that, the programme consistently went on the road, including two visits to the University of Manchester, to discuss 'What's the North Ever Done for Us?'

and 'Science and Spin'. With over 170 programmes – still produced by Feachem – the series has won major international awards, has global reach through the podcast and is firmly fixed in the BBC radio schedule. Above all, it represented the realisation of a key aim of Cox and Milinovich – the fusion of science and popular culture.

Government initiatives

Encouraged by and further stimulating this media activity, public interest in science grew – Manchester launched its annual Science Festival in 2007, with the enthusiastic involvement of researchers from the University. Government departments and research funders were also keen to change public attitudes to science, so they developed initiatives to encourage researchers to communicate their findings, such as six regional 'Beacons for Public Engagement', based in Manchester, Newcastle, London, Norwich, Edinburgh and Cardiff.[10] Supported by over £9 million of funding from Research Councils UK and the Wellcome Trust, the Beacons spread best practice for how to engage the public, and in particular how to estimate if all this activity had any consequence – was there any lasting change in public understanding, attitudes or behaviour?[11]

In Manchester, the Beacons were warmly welcomed. Nancy Rothwell, who had a record of explaining her scientific research to the public, including through the BBC's Royal Institution Christmas Lectures, was at this time the University's Vice-President for Research. As the principal investigator of the Manchester Beacon, she encouraged researchers to use this new funding:

> this new initiative will allow us to integrate ourselves into local communities and discover what people really want to know about our work [...] This is a chance for us to learn from people across the Manchester area, to build bridges, make universities integral to the wider community and seek opportunities to make an impact through engagement.[12]

Part of the aim of the Beacons was to encourage a shift from the traditional lecture-based, scientist-as-expert approach towards novel and exciting ways of engaging and interacting with the public. In

Manchester this took the form of regular activities at the University's Manchester Museum – a popular choice with families at weekends – and in University buildings, beginning with a Community Open Day in the Faculty of Life Sciences, focused on the communities living on either side of Oxford Road, which eventually became an annual University-wide Community Festival. These activities would involve stalls, activities, displays and talks aimed at every age group. One notable initiative was the Brain Bus, led by neuroscientist Stuart Allan, which visited railway stations, secondary schools and public libraries. All this activity was coordinated within the University under the rubric of social responsibility, a unique feature of the institution's mission that was developed and supported by Nancy Rothwell after she was appointed vice-chancellor following the retirement due to ill health of Alan Gilbert in 2010.

At a national level, the significance of public engagement was codified in 2011 by the Concordat for Engaging the Public with Research, signed by higher education and research institutions, research funders and government, which led to the creation of a network of 'public engagement champions' in UK universities to encourage and coordinate engagement with research, and particularly with science.[13] Further encouragement of engagement activities was just around the corner, as a key word used by Rothwell – 'impact' – became omnipresent in UK academia. Underlying both the Beacons and the Concordat was a realisation in government and the academic community that the public needed to be convinced of the importance of spending money on research, in particular following the financial crash of 2007–08 and the subsequent wave of austerity and public expenditure cuts. The UK Research Councils and government therefore sought to find proof that scientific research was not simply exciting and intriguing but also that it had real consequences for people's lives – 'impact'.

This term was introduced into the UK Research Excellence Framework (REF) in 2014. REF is a septennial exercise that determines funding to UK universities on the basis of their research activity. Most of the REF funding is allocated according to the perceived excellence of the publications produced by each institution, but from

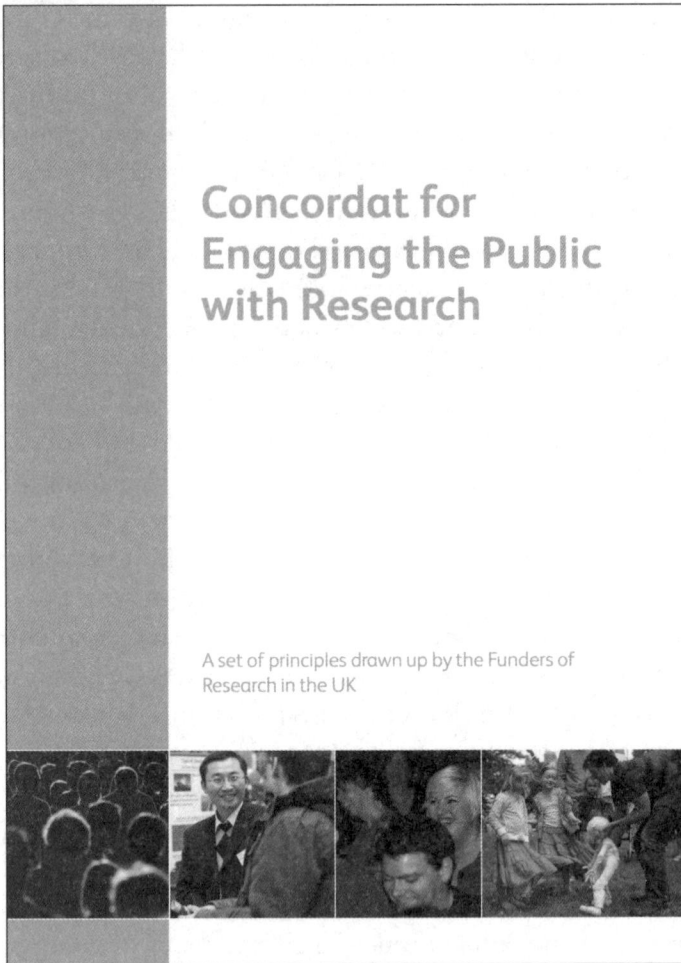

15.1 The Concordat for Engaging the Public with Research (2011).

2014 a proportion was devoted to 'impact' – 'an effect on, change or benefit to the economy, society, culture, public policy or services, health, the environment or quality of life, beyond academia'. Public engagement with, and understanding of, science was seen as one key part of impact. Not only was such activity stimulating and fun, it could now bring lots of money into the universities. Alongside a change in the public's attitude to science, scientists' attitudes to the public changed as university administrators recognised the literal and metaphorical value of such work and encouraged and supported it.

The *Wonders* years

2010 saw yet another significant historical celebration – the 350th anniversary of the foundation of the Royal Society of London, the world's longest-running scientific institution. Ever attentive to the power of an anniversary, the BBC decided to devote a major part of its output in 2010 to science, and branded it the Year of Science.[14] Among the many media events of that year, Martin Rees, the outgoing president of the Royal Society, gave the Reith Lectures, the plasticine duo Wallace and Gromit featured in a six-part children's series about inventions, and Brian Cox presented *Wonders of the Solar System*. This was the series that thrust him into the consciousness of the British public, and also that of 10 Downing Street – he received an OBE in 2010 for services to science.

Cox's five-part series, with Andrew Cohen as executive producer and Bradshaw and Olding writing and directing, was broadcast on BBC 2 in March–April 2010 and saw Cox visiting various exotic locations to convey the nature of the sun and the planets, showing the explanatory power of physics. It also demonstrated the continuing value of one-way traditional science communication: the series consistently attracted audiences of around three million and went on to win a prestigious Peabody Award in the USA the following year, as well as awards for Cox in the UK. A book of the same name, co-authored with Andrew Cohen, sold over 100,000 copies. This was impact, and indeed was later cited as part of the University's 2014 REF impact case study relating to the particle physics research carried out by Cox and his friend and colleague Jeff Forshaw.

That year the *Guardian* ran a feature article entitled 'How Science Became Cool'. Ten personalities, from comedians to science writers to researchers, were asked the question 'Is this a golden age of science?' The article was illustrated, inevitably, by a photo of Cox, dressed in mountain gear and seated on the Matanuska Glacier in Alaska. As Cox's ubiquity indicated, it was certainly a golden age for science broadcasting, with the BBC commissioning science programmes from presenters such as the theoretical physicist Jim Al-Khalili of the University of Sussex, the geologist Iain Stewart of the University of Plymouth and the anatomist Alice Roberts, then of the University of Bristol.

The massive success of *Wonders of the Solar System* was not the only boost to physics in the UK and in Manchester that year – in October, University of Manchester researchers Andre Geim and Konstantin Novoselov won the Nobel Prize in Physics for their work on graphene. This led to a wave of construction and investment in Manchester as attempts were made to develop applications for the new wonder substance. It also propelled materials science to a leading position in the University's research portfolio, culminating in the 2022 opening of the Henry Royce Institute, the UK's national institute for advanced materials research and innovation, on the University campus.

2011 saw Cox presenting a four-part follow-on series, *Wonders of the Universe*, in which his globe-trotting style was used to explain key features of the universe, particularly time and gravity. The first episode, which aired on BBC 2 in March 2011, was marked by Cox's first appearance on the cover of the *Radio Times* ('The hottest star on TV') and was seen by over six million people; again, there was a book tie-in of the same title, co-authored by Cohen.[15]

Cox was now well known to the British public. At the same time there was a substantial increase in the number of students applying to do physics at British universities, and in particular at Manchester. Although the causes of this shift were complex, the media latched on to 'the Brian Cox effect' and used the term to explain the growth in university science applications and the public's interest in science. This undoubtedly helped recruitment to the Manchester physics course – it rapidly became the most popular in the country and the most difficult to get into. Those lucky and talented students who gain a place are not short-changed – Cox still teaches an introductory course on quantum mechanics.

This initial effect has been reinforced by Cox's subsequent media work, which now has a global reach, including several television series, such as *Wonders of Life* (2013), *Human Universe* (2014) and *The Planets* (2019), many best-selling books, as well as extraordinary sell-out stadium tours around the world, in which he explains particle physics to audiences of tens of thousands. However, although Cox's Manchester identity was always front and centre – not least because of his Oldham accent and the prominent credits to both the University and various colleagues who acted as advisors – the link between many of these programmes and the University of Manchester was not always obvious. One clear exception was his 2014 series on the history of British science and the importance of supporting curiosity-driven research, *Science Britannica*, in which Cox described Manchester as 'the greatest university in the history of the universe'.

The direct link between Cox's media work and the University became evident in January 2011, when a live TV mini-series, *Stargazing Live*, was launched on BBC 2. Fronted by Cox and Irish comedian Dara Ó Briain (who has a degree in mathematics and theoretical physics) and broadcast on three successive evenings, this programme about astronomy was broadcast from the University's radio observatory at Jodrell Bank. The visuals were dominated by the iconic Lovell telescope, which was dramatically lit up by searchlights. *Stargazing Live* sought to do for astronomy what *Springwatch* and *Autumnwatch* – similar live BBC TV factual programmes – had done for the natural world. Members of the University featured prominently, in particular the associate director of Jodrell Bank, Tim O'Brien, who was the local host. Indeed, an early version of the *Stargazing* idea had been broadcast from Jodrell in August 2003 – a late-night live programme, *The All Night Star Party*, with O'Brien as the host.[16] *Stargazing Live*, which aired from Jodrell every January between 2011 and 2016, attracting around three million viewers for each programme, not only helped reignite public recognition of Jodrell Bank and the University, but also saw the beginning of a new, interactive form of public engagement with science – citizen science.

An inevitable feature of broadcast media is that it tends to be one-way – the presenter talks to the public in as engaging a fashion as possible, hoping that the viewers and listeners will understand and be inspired. But the wave of activities encouraged by the Beacons, supported by research by social scientists from around the world, suggested that interactive forms of engagement were more productive for both public and scientists. By involving people directly in research, new enthusiasms could be sparked and researchers could gain insight from the ideas and activities of the public. Rather than a one-way street, public engagement, or science communication as it increasingly became known, could become a circular, progressive affair, enriching all participants.

Stargazing Live exemplified this approach by showing viewers how to use a telescope (sales soared) and encouraging people to join local astronomy groups (the Macclesfield Astronomical Society featured strongly) and, in the second series, by introducing the public to Galaxy Zoo. Created in 2007, this was the brainchild of broadcaster and astronomer Chris Lintott and astrophysicist Kevin Schawinski, both of the University of Oxford; it was a website where members of the public were invited to classify images that computers could not accurately process. By using the power of the crowd, Galaxy Zoo projects were able to screen through millions of images, identifying the presence of exoplanets and classifying hundreds of thousands of new galaxies. On the 2014 show, discoveries made by the public live on TV led to a publication, with Cox and O'Brien among the co-authors.[17] Soon expanded beyond astronomy to biology and the humanities, under the umbrella title Zooniverse, this ecosystem has millions of users and has enabled ordinary citizens to participate in research, leading to many academic publications – over sixty through the Galaxy Zoo projects alone.

From *Stargazing Live* to bluedot

Stargazing Live reinforced the recognition of the Jodrell Bank observatory, which has been present in British popular culture since its 1957 detection of the first artificial satellite, Sputnik. For example, Jodrell featured in a 1980 *Doctor Who* episode, watched by six million viewers, when the fourth Doctor (played by Tom Baker) tussled with his nemesis The Master, fell off the Lovell telescope and regenerated into the fifth (played by Peter Davison). Jodrell has also featured in various pop music videos, including one by Brian Cox's band, D:Ream. The clearest recognition of the significance of Jodrell Bank was its inscription as a UNESCO World Heritage Site in 2019 (this was the result of a decade of work led by Teresa Anderson).

At the end of the twentieth century the facilities for the public at the Jodrell site were poor and the handful of public buildings were eventually demolished. Following the arrival of Teresa Anderson at Jodrell in 2006, greater attention was paid to public engagement, and in 2011 a new Discovery Centre was opened – between 2013 and 2021 over one million visitors came to the site.[18] As well as traditional outreach activities such as lectures and displays, the new centre organised female-only 'Girls Night Out' events to encourage women and girls to be interested in astronomy and physics. This work culminated in the opening of an impressive new visitor building, First Light, which cost £23 million and was supported by the Heritage Lottery Fund and the UK government.

In 2011 an initiative led by Anderson and O'Brien saw the first public musical event at Jodrell, with a concert by the rock band The Flaming Lips in front of the Lovell telescope, which was used as a giant projection screen for a light show. The 5,000 visitors not only enjoyed the music, they also thronged to hear talks on science that were presented before the concert started. Such was the interest from the public that the idea was repeated in the following two years under the title Live from Jodrell Bank, with a blend of music that sometimes had a Mancunian tinge (New Order and the Hallé Orchestra both performed) as well as scientific talks, stands and displays.

There were precedents for this kind of event. For example, since 2008 Robin Ince had hosted an annual Christmas show in London, with a blend of science, music and comedy (in 2009 it was co-hosted by Brian Cox and broadcast on BBC 4). But the Jodrell events were on a different scale and in 2016 the first bluedot festival was held over three days, with techno/spacey music (Underworld, Jean-Michel

Jarre and others) and science talks and stands, as well as a broadcast by *The Infinite Monkey Cage* and the creation of science-related artworks. Held every year between 2016 and 2019, the event was attended by over 70,000 visitors and attracted huge media interest and several awards. After a hiatus due to the COVID-19 pandemic, the festival returned in 2022 and 2023, with similar success (plate 7). The *Guardian* reported that bluedot is 'a unique festival in which science and music go hand-in-hand, and where homemade spacesuits are as plentiful as band T-shirts'.[19]

In many ways, bluedot – a unique event, unequalled by any university anywhere in the world (or the universe) – represents the culmination of all the initiatives and experiments in communicating science and engaging the public with research that began to bubble up in the early years of the century. Audience surveys have consistently revealed people's enthusiasm for being able to discuss directly with scientists. In return, scientists are inspired by the public and the response their work received.

Conclusion

Over the first two decades of the twenty-first century, an astonishing amount of enthusiasm and imagination were deployed by scientists, many from the University of Manchester, seeking new approaches that could inform and inspire the public. In the shape of Brian Cox, the British and eventually the global public found someone who could explain complex ideas in simple and inspiring ways.

This period of innovation undoubtedly changed public attitudes to physics and to science in general. But many sectors of the British public are still not encountering these initiatives, in particular people in more deprived regions and members of ethnic and social groups that are not, for example, likely to attend bluedot or listen to Radio 4. Furthermore, the pandemic revealed – and probably fomented – anti-scientific views among many sections of the population regarding vaccination and even the existence of the coronavirus. These views were not limited to rejecting scientific understanding, they sometimes spilled over into alarming hostility towards scientists who were communicating their knowledge and were perceived by some of their opponents as corrupt or manipulated by shadowy forces. Even physics has been affected, as shown by the bizarre recurrence of belief in a flat earth.

These developments show that, even with such diverse activities around science, its methods and its ideas, communicating science remains a labour of Sisyphus, as new developments prompt both an embracing of scientific explanations but also a recurrence of anti-scientific ideas and prejudices. In the clearest fashion possible, these decades show that science is not outside general culture, but is fully part of it, and that attitudes to it can shift both through the activities and initiatives of scientists, but also through the challenge of events.

*   *   *

My thanks to Teresa Anderson, Brian Cox, Alexandra Feachem, Robin Ince, Gia Milinovich and Tim O'Brien, who kindly gave up time to be interviewed for this chapter.

Notes

1 Brian Cox tweeted 'Anyone who thinks the LHC will destroy the world is a twat.'
2 *Radio Times*, 4 September 2008, schedule for 6 September, https://genome.ch.bbc.co.uk/issues (accessed 28 March 2024).
3 'Putting the fizz into physics', *Observer*, 14 September 2008, https://www.theguardian.com/science/2008/sep/14/cern.particlephysics (accessed 28 March 2024).
4 *Radio Times*, 30 December 2004 and 13 January 2005, schedules for 1 January and 15 January 2005.
5 *Radio Times*, 26 April 2007, schedule for 28 April.
6 https://web.archive.org/web/20230610053450/https://cernpodcast.wordpress.com/ (accessed 28 March 2024).
7 https://web.archive.org/web/20230528003935/https://cernpodcast.wordpress.com/2007/11/27/science-and-religion/#comments (accessed 28 March 2024).
8 *Radio Times*, 24 January 2008, schedule for 26 January.
9 Particle Physics Impact Case Study, UoA9 Physics, REF2014, https://web.archive.org/web/20231125101734/https://ref2014impact.azurewebsites.net/casestudies2/refservice.svc/GetCaseStudyPDF/28175 (accessed 28 March 2024).

10 University of Manchester press release, 9 November 2007, https://web.archive.org/web/20231125100917/https://www.manchester.ac.uk/discover/news/manchester-beacon-will-use-engagement-to-build-bridges-with-local-communities/ (accessed 28 March 2024); Sophie Duncan and Paul Manners, 'Embedding public engagement within higher education: lessons from the Beacons for Public Engagement in the United Kingdom', in Lorraine McIlrath, Ann Lyons and Ronaldo Munck (eds), *Higher Education and Civic Engagement: Comparative Perspectives* (New York: Palgrave Macmillan, 2012), pp. 221–40, doi: 10.1057/9781137074829_14.

11 Research Councils UK, *What's in it for me? The benefits of public engagement for researchers*, https://web.archive.org/web/20230803201551/https://www.ukri.org/wp-content/uploads/2020/10/UKRI-16102020-Benefits-of-public-engagement.pdf (accessed 28 March 2024).

12 University of Manchester press release, 9 November 2007.

13 Funders of Research in the UK (2011), *Concordat for Engaging the Public with Research*, https://web.archive.org/web/20231013001830/https://www.ukri.org/wp-content/uploads/2020/10/UKRI-151020-ConcordatforEngagingthePublicwithResearch.pdf (accessed 28 March 2024).

14 *Ariel*, 19 January 2010.

15 *Radio Times*, 5 March 2011, front cover.

16 *Radio Times*, 21 August 2003, schedule for 23 August.

17 J. E. Geach et al., 'The Red Radio Ring: a gravitationally lensed hyperluminous infrared radio galaxy at $z = 2.553$ discovered through the citizen science project SPACE WARPS', *Monthly Notices of the Royal Astronomical Society*, 452 (2015), 502–10.

18 Jodrell Bank Impact Case Study, UoA9 Physics, REF2014 and REF2021, https://web.archive.org/web/20231125133747/https://ref2014impact.azurewebsites.net/casestudies2/refservice.svc/GetCaseStudyPDF/28173 and https://web.archive.org/web/20231125133249/https://results2021.ref.ac.uk/impact/12f1cecc-b557-4506-85e2-4b63820829fc?page=1 (accessed 28 March 2024).

19 'Bluedot festival review', *Guardian*, 25 July 2022, https://www.theguardian.com/music/2022/jul/25/bluedot-festival-review-strings-and-stars-jodrell-bank-observatory-bjork (accessed 28 March 2024).

16

Post-crash economics: liberal education through struggle against the curriculum

Joe Earle, Sukhdev Johal and Karel Williams

> It really was a life changing experience and I don't think many people can say that about the societies they join at university, and I think many others would feel the same.
>
> (question 21, respondent 22, on being a member of the Post-Crash Economics Society)

The idea of self-developing, character-forming liberal education has been around for more than two hundred years, and the *Oxford English Dictionary* gives usage quotes going back to the 1700s. John Henry Newman in the mid-nineteenth century connected this established concept with 'the idea of a university' as the institution which through curriculum and pedagogic relations could deliver the necessary formation of character.[1] Ideas about the appropriate university curriculum have changed radically with the times. In a secular society few would give theology a privileged place as Newman did, while the corporate universities of our time might claim that all the 'disciplines' of humanities, social sciences and natural sciences can be and are taught in ways that deliver a liberal education.

The concept of liberal education has long been associated with a binary distinction between academic and vocational knowledge. Thus, the *Dictionary of Education* distinguishes between the qualities of reasoning and critical reflection acquired by studying abstract or interpretative knowledge as an end in itself, and the technical, instrumental knowledge necessary for a practice.[2] Academic knowledge (as dispensed by universities) is then conventionally presented as the higher form of self-development, which has great value for those who are critical contributors to a diverse, changing and complex society. This is so whether they contribute as managers taking decisions that shape our lives or as citizens equipped to participate in what Nussbaum calls 'a critical public culture'.[3]

All of this conventional thinking is contested. The criticism is that what now passes for liberal education encourages modernist technocratic fantasies of top-down policy control, while at the same time it can also spread the practice of an amateur ruling class. Thus, James C. Scott classically criticised state and corporate reliance on thin simplifications of knowledge at a distance and inverted the high/low order of worth by arguing the superiority of mētis, which is a contextual, practical, flexible knowledge of practice.[4] Meanwhile, amid mounting criticism of British elites, Simon Kuper argues that the Oxford Tories who took over Westminster and the media after New Labour's defeat had been formed by humanities degrees and Union debating which taught them to 'write and speak for a living without much knowledge'.[5]

Thus, liberal education for its advocates should be about high ideals and serving the public good, while for its critics it is about encouraging the damaging illusions of our officer class in economy and polity. It was against this background that in late 2012 Manchester students launched the Post-Crash Economics Society, whose central demand was for pluralist reform of the existing economics curriculum that focused on induction into the one mainstream paradigm. Their implicit contention was that the existing syllabus was not fit for the liberal purpose of equipping students to serve as contributors, whether as managers or citizens. The first section of this chapter presents a narrative account based on first-hand understanding of events and outcomes as academic economists and university managers adjusted the curriculum without conceding pluralism. The second section tells a different story about the value of the experience based on a questionnaire returned in 2023 by ex-student activists. As in the headline quote above, they all report that the experience of being part of Post-Crash was a valuable and formative one. These students gained their liberal education not through the curriculum but through their struggle against the curriculum.

A narrative of events and outcomes

The story of the University of Manchester's Economics Department and its relations with students in the 2010s and after can be told in many ways, highlighting different events and outcomes. University managers might highlight the successful efforts of academic staff responsible for teaching and learning in school and department to improve scores in the National Student Survey in line with the goals of the University leadership. This chapter offers a bottom-up, alternative account that focuses on organised student demands for more pluralism in economics education, and the disappointing outcome of a struggle that achieved modest reform but no kind of pluralism. This narrative of frustrated demands frames the student experience reported in the second half of this chapter.

This account is largely based on the authors' first-hand experience of events. Earle as a student was one of the founders of Post-Crash and was subsequently a co-author of the 2017 book which presents the founders' collective account.[6] In the early years of Post-Crash, Williams was a director of the Economic and Social Research Council funded Centre for Research on Socio Cultural Change at the University of Manchester and a senior academic supporter. Johal, based at Queen Mary University of London, researches with Williams, and both are long-standing internal critics of the governance of corporatised universities through mechanisms such as research assessment. It should be emphasised that the authors' account of events and outcomes in the Economics Department at the University of Manchester is entirely coherent with peer-reviewed accounts of system-wide developments in research-led economics departments across the UK. There is debate about how to understand the mechanisms of change, as between Whitley and Lee et al.[7] But there is widespread agreement that system-wide research assessment in economics narrowed the definition of quality research, and through hiring and promotion decisions in research departments of economics made any kind of pluralism increasingly difficult. The articles by Lee and Stockhammer et al. present overwhelming evidence on this point.[8]

Post-Crash Economics began as a society founded by students of the Economics Department at the University of Manchester in late 2012. It grew out of initially unfocused discontent about the economics curriculum. As the name of the society indicates, one important initial issue was that the departmental curriculum included very little that would help students understand the causes of the 2008 financial crisis, its consequences, and the subsequent struggle over regulation of the finance sector. A first email to students in October 2012 summoned 'econosceptics' and just five students turned up. Six months later 200 students came to the launch event in March 2013.

By this point, students had formulated broad demands for more pluralism in the economics curriculum for good liberal reasons. Their petition to the department complained of four absences from the curriculum: lack of alternative non-mainstream perspectives, lack of real-world applications, lack of historical perspective and a general lack of critical thinking. This agenda was reflected not only in the student demands but also in the self-education events that the society sponsored. These included a lecture series with a variety of speakers under the title 'What You Won't Learn in an Economics Degree' and an evening not-for-credit course, 'Bubbles Panics and Crashes', which introduced different perspectives on economic crises. In the title of their launch event, Post-Crash asked 'are economics graduates fit for purpose' (as managers and citizens in line with the ideals of liberal education) (Plate 8)?

Some outsiders saw this as leftist rebellion. But the students always insisted that their liberal agenda was radical but not of the left or right. It was certainly not party political when society activism could be understood as an alternative to traditional forms of student involvement in the youth wings of political parties or student unions. In terms of economics, the students were as interested in Hayek and the Austrian School (individualistic and market-based) as they were in Keynes and the varieties of Keynesianism. As they explained in a press pack,

> Our society is not about left-wing or right-wing politics or economics; it is about the dire state of national economics education today... We firmly believe that Hayek as well as Keynes would both be turning in their graves if they knew what an undergraduate economics education consisted of today. More importantly, we believe that our economics education system is not able to produce economists of the calibre of Keynes or Hayek anymore and it is society that will pay the price. Keynes and Hayek were both well versed in a broad range of economic thought and history, they both had vastly developed critical capabilities, they were great writers and communicators as well as economists and, most importantly, they were both free thinkers who overturned existing ideas and norms to build their respective theories. All that is now lost...[9]

This liberal analysis divided the Economics Department. There was strong support for student demands from some junior staff, but only a small number of senior staff in the department were supportive; many of the senior staff who aligned with the students were to be found in other departments such as Politics or the Business School. The managers of the Economics Department (head of department and research director) appeared to the students as not so much hostile as uncomprehending. At the Post-Crash launch event in March 2013, the head of department publicly compared alternative perspectives in economics to the outdated practice of tobacco smoke enemas in medicine.[10] When the staff member offering the 'Bubbles, Panics and Crashes' course did not have his temporary contract renewed, 'budgetary constraints' prevented the department from running the course.[11]

The academic rationale for defending the existing mainstream curriculum was straightforward. The Manchester economics curriculum on micro- and macroeconomics, econometrics and statistics was much the same as in peer institutions in the UK and elsewhere, which all used a few standard textbooks. This curriculum represented the consolidation of economic knowledge that students needed to know; what they were not currently being taught was of little value and interest. The implication was that it was not necessary for economics students to know the history of economic thought because anything of value in the discourse's history had been incorporated into current theory, which systematised knowledge in ways that made economic history otiose. Events such as the Great Financial Crisis could be dealt with by introducing a financial sector into macro models. Behavioural economics might have small-scale experimental proof that subjects did not consistently make calculations of marginal utility, but that did not require a wholesale reworking of microeconomics.

The department whose managers took this position was an invention of recent date, because twenty years previously it had been a lively pluralist department with a diverse range of offerings. In the early 1990s the Economics Department included an active development economics group, and it worked alongside an Agricultural Economics Department which was formally independent. Both units incubated a good deal of heterodoxy, because the history of development economics is about paradigm disputes over how to understand the process of development, while, in agricultural economics, family farms work in all kinds of ways that are difficult to fit into standard models of the firm. The departmental stars in the mid-1990s included Diane Elson, a leading feminist economist attached to the development economics unit. Other staff in the Economics Department included Pat Devine, a Marxist with a practical interest in industrial economics and participatory planning, and Terry Peach, a historian of economic thought with publications on Smith and Ricardo.

By the mid-2000s much of this diversity had been lost as such researchers were fewer in number and their diverse viewpoints were not represented in the mainstream curriculum. The driver of this process was external assessment of research quality, which drove a redefinition of quality and what counted in economics. The first Research Assessment Exercise took place in 1986; it was subsequently repeated typically at 4–5 year intervals and was renamed the Research Excellence Framework (REF) in 2014. Its double purpose has always been to provide some accountability for past funding and to influence the distribution of future funding. In all assessments the grading is of units of assessment (which did not always correspond with departmental boundaries), with weight given to individual publications in various ways. In each unit of assessment, a peer review panel grades the research of individual units according to the same status hierarchy. Top grades indicate world-class quality, and the grading then moves downwards through various kinds of international recognition towards national recognition. This hierarchy is manifestly most relevant to large-grant, big-team, hard natural sciences, but it has been applied to humanities and social science disciplines.

The interesting point is that the same method and hierarchy produced very different outcomes in different social sciences. In politics and sociology it licensed pluralism, while in economics it empowered the mainstream to evict the heterodox, because in economics quality was increasingly identified with papers published in a few highly ranked mainstream journals. The politics and sociology departments at Manchester and other research universities contained diverse research groups whose theoretical assumptions and substantive concerns were very different; a departmental REF entry was, then, a portfolio with the achievements of all the groups talked up. New hires would be distributed between various groups to plug gaps as and where necessary. A strong departmental entry in economics was a list of heavily quantitative articles published in high-status mainstream journals which would reject heterodox submissions as outside the scope of their journal. New hires would be made insofar as they were likely to publish in the right journals, whose editorial boards were dominated by mainstream economists. The end result is that, as a quantitative study of journal rankings and economics submissions in REF 2014 concluded, 'the REF in its present form marginalises heterodox economics, pushes it out of the economics discipline and endangers pluralism in economics research'.[12]

By the mid-2010s, in a parallel process, Manchester University (like other UK universities) had been corporatised and strategised so that the School of Social Sciences and the Faculty of Humanities played a much greater role in managing departments. This was the context in which the students produced their 2014 report, *Economics, Education and Unlearning*, which analysed the Manchester undergraduate curriculum in detail.[13] The students then met with senior managers at school and faculty levels without obtaining any positive result. As they afterwards complained 'we felt we were being passed around different people ... without making any substantive progress'.[14] School and faculty managers were held accountable for their units' performance against key metrics, and REF scores were important performance indicators. The role of senior managers was to restructure and improve units with low REF scores, to reward those with high REF scores, and to live uneasily with those ranked as middling. It was not managers' role to question what drove grades or whether there was an inverse relationship between REF success and curricular breadth.

The publications-based REF status of individuals was a central determinant of hiring, tenure and internal promotion decisions. Those staffing decisions in turn shaped the kind of curriculum that could be offered. The narrow definition of what might be called 'REFability' in economics created a kind of internal institutional alliance against pluralism in economics (at department, school and faculty level), which held up against the external pressures for change from below and outside the University. These pressures were, however, magnified in two ways.[15] First, Post-Crash students ran a brilliant national media campaign on the back of their petition. Their report on the Manchester curriculum had a supportive foreword by Andy Haldane, then chief economist at the Bank of England. Their *Guardian* newspaper article of October 2013 was downloaded on Facebook 26,627 times. Their May 2014 report on the Manchester curriculum was downloaded more than 15,000 times in six months.[16] Second, the Manchester students linked up with other dissident student groups that were spontaneously appearing in the UK and abroad. The result was the formation of the Rethinking Economics network for students in the UK and the rest of the world. By May 2014 a global petition calling for pluralism had been signed by 63 groups from 230 countries.

The level of system-wide student discontent was such that mainstream academics in Manchester and other research-based economics departments had to make some concessions. Manchester hired economics journalist and consultant Diane Coyle as a visiting professor to teach a module on public policy, added a module on economic history and made some changes in core modules. At a national level, Wendy Carlin of UCL led the CORE initiative (Curriculum Open-Access Resources in Economics), which produced a now widely used introductory textbook that added some economic history and relevant examples. CORE came with a wrapper of 1960s positivism about 'testing' because the hope was that economists 'will learn to use evidence from history, experiments and other data sources to test competing explanations and policies'.[17] This was movement in the right direction but still much less than the students had requested at Manchester. In the CORE syllabus, there was no concession to the central demand for pluralist teaching of, and respect for, different economics paradigms.

The experience of student activists

Our account of the student experience is based on a small questionnaire survey of those who had been active members. Of the activists who were sent the questionnaire, 23 returned it. Their returns were anonymised but gave us basic information on background, current employment and years when actively involved in the society. The 23 respondents divided more or less evenly between 11 who were founder/first-generation members of 2012–14 and 12 who joined in subsequent years. The society from the beginning had a committee but the role division was informal, especially in the early years. Thus, three respondents struggled to remember what their formal responsibilities had been. But all vividly remembered their self-education and organising activities: publishing blogs, taking part in reading groups, organising events, giving media interviews, representing the society in meetings with senior staff.

These economics students came from varied social backgrounds and from many different places. One had gone to a 'bog standard comprehensive' in a 'challenging area', while another had been mentored by a grandfather who had studied under Hayek at the London School of Economics. The reputation of Manchester as a city and the University as a member of the Russell Group did attract applicants, but there was also an element of happenstance because many could have gone elsewhere if A level grades or admissions tutor decisions had been different.

After graduation the members dispersed into a variety of occupations in ways which disprove the stereotype that the majority of Post-Crash activists were leftist troublemakers. Most of the activists became what one respondent termed 'radicals in suits'. Another observed that Post-Crash's 'reach beyond the universities' had been achieved by students 'moving into influential jobs'.

Career destinations are interesting because, despite often excellent exam results, only two student activists became academics and just one managed to get into the downsizing profession of journalism. There were three main career destinations:

- Six of the activists are now employed in staff roles as policy advisors or staff economists in government. Several work in Whitehall ministries and the Treasury while one works in a local authority. Two mainland European students have roles in EU policy development and ECB banking supervision.
- Five of the activists are now employed in a variety of roles in the mainstream of the finance sector. They occupy a variety of mid-career, mid-level roles. One works at a trading desk, another in venture capital and a third in risk management. Two more are vice presidents, leading departments in one of the world's largest fund investors.
- Four of the activists work for NGOs. This is where the leftist minority have ended up because three of these NGO employees work for small radical NGOs. But even here the mainstream is relevant because a fourth ex-student now has responsibility for a major policy area in a centrist think tank which operates explicitly in a cross-party way.

Most respondents had open motives for studying economics, which they hoped would answer big questions. One wanted to understand 'how the economy works', another to understand 'how the world works'. Initial interest in the society was sometimes a matter of 'curiosity' or 'socialising'. In response to a question about what they were hoping to achieve when initially joining the society, one reply was 'nothing, just seemed cool'. But for most respondents, the active driver and agent of radicalisation was student discontent with the economics syllabus of the department, where 'economics seemed divorced from the real world and all the big questions that led me to the subject in the first place' (question 5, respondent 19). This was particularly an issue for students on the PPE degree scheme who could contrast the style and content of teaching in politics and philosophy with that in economics; but it was also raised by specialists on the BA Econ degree.

> Having come to university to understand the world better I was feeling frustrated, disenfranchised and disillusioned that none of the major issues of our time (climate change, financial crisis, inequality, poverty) or anything that I saw on the news was being explained or even mentioned in my lectures. I was aware that there was a lot more to economics than the IS/LM model, but I knew that what was in our textbooks had precious little resemblance to what was going on in the world.
>
> (PPE undergraduate, question 5, respondent 7)

> From studying A level economics, I was already disappointed about the lack of different perspectives and when it got to degree level the amount of real-world application reduced further and there seemed to be no proper discussion of economic inequality in the course.
>
> (BA Econ undergraduate, question 5, respondent 2)

As already noted, the students campaigned for the one central objective of pluralism, that is, that the department should teach and respect different economic paradigms. Their supplementary demands were for more real-world examples and economic history with less emphasis throughout the curriculum on multiple choice questions and more on traditional liberal arts skills of essay writing and discussion. In retrospect, they were all clear that in the University of Manchester and other economics departments, the academics made concessions on the supplementary demands without conceding pluralism. This was very clear in their response to the CORE syllabus and textbook, which represented the system-wide response of progressive mainstream UK economists. One respondent conceded that

CORE was 'a substantially better product'. But the consensus was that radical change had been bought off by tactical concessions because CORE 'addresses real world application and economic history but not the (pluralism) issue'.

> Hierarchies work in clever ways. When CORE economics was introduced, there were lots of people who were happy as it meant a new way of doing economics. But ultimately when you read through the materials it was like putting lipstick on a pig. The mainstream had taken the problem and only listened to some of it (we need more economic history and real-world examples); they then claimed they had listened.
>
> (question 10, respondent 4)

> The CORE syllabus ... I see as a case of changing just enough to stay the same even though I think its authors genuinely believe they are significantly reforming economics education... They certainly didn't embrace a pluralist education where economic theory is contested. They still taught from the position of a single coherent pretty narrow body of knowledge that is economics while other perspectives are bad economics or not actually economics at all. In that goal we haven't yet succeeded.
>
> (question 11, respondent 23)

And yet all respondents believed that despite losing on pluralism, the struggle had been a hugely valuable 'learning experience' and all agreed the learning was 'not just (about) different perspectives on economics and politics but about organisation and campaigning from a more practical perspective':

> The main way it changed me was the process of meeting a group of young like-minded people who are interested in ideas, politics, economics and crucially doing something about it. I had always hoped I would meet people like this in life and it felt very special to meet them at Manchester.
>
> (question 9, respondent 5)

The students learnt about power and how institutions work to manage dissent and conservatively limit change. Many students started their campaign by 'naively' thinking that there was 'an open market in ideas' so that 'a few meetings and some press coverage would put enough pressure on the lecturers'. But they graduated with a completely different understanding of institutional politics:

I saw up close how institutions avoid change and respond to challenge. How responsibility is passed to different places so it's very hard to pin down. How calls for change can be channelled into reviews and reports. How smaller reforms are used to head off bigger changes. How change efforts can be branded as something positive for the institution. How the business model, wider research funding and decision-making structures of U of M reproduced the education we were trying to challenge and made the scope for change much smaller. And how the academic debate about the merits or otherwise of curriculum reform paled into insignificance in terms of its influence on whether and what change happened.

(question 10, respondent 23)

This understanding of 'how universities really worked' did not lead to any kind of hostility towards individual academics or university managers. One respondent 'wished I had been more sympathetic to the department/lecturers and their reluctance to change'. The problem with most mainstream economists was sunk cost and 'the serious intellectual investment that individual academics had in their ideas'. More generally the problem was system constraints on leadership in the department and school of social sciences:

I felt the leadership fell into two camps: those that agreed with us but couldn't do anything due to external constraints and those that just disagreed with us. On the former, it taught me that university managers are operating in a highly constrained environment driven by external factors such as the REF and were responding to those constraints in a rational way. In essence I could understand why they didn't want to do what we were telling them. On the latter I sensed that there were those in the leadership that simply thought we were crazy and unscientific and trying to undermine the enormous progress economics had made to become the premier social science discipline.

(question 10, respondent 5)

In this context, the objective of pluralism could never be achieved, but the struggle against the department for pluralism and the performance of pluralism within Post-Crash was hugely valuable:

Challenging authority with a view that's not popular to begin with and persisting despite setbacks is a hugely transformational experience. I am so much more confident and back myself to challenge others/

say the unpopular thing sometimes. These experiences helped me get the job I have now and make me much better at it.

(question 9, respondent 9)

Within the society's committee, and across committees over time, there was a lot of variation in opinions on the role/value/ambitions of the society. I think the value comes exactly from this debate around current economic and social issues and that the value of Post-Crash comes from organising events and debates in parallel to the department's teaching.

(question 21, respondent 20)

One question asked the activists whether the campaign 'influenced what you did next and what you have done since'. Out of 23 respondents, 22 believed that Post-Crash activism had influenced their choice of subsequent career and career development, from joining the civil service to bond trading. As one respondent replied, 'I have used my campaigning skills and economic knowledge in all my subsequent roles.' The campaign, more so than any economic knowledge, had been a liberal education that prepared them for a long march through the institutions because it was 'the most formative and enjoyable experience I had at University'. As another wrote, 'I was 100% changed by my discussions in Post-Crash and by my engagement with the economics department. I remain inspired to this day.'

It is therefore encouraging that more than ten years after its foundation, Post-Crash Economics continues to operate as a student society at the University of Manchester with the support of an international organisation, Rethinking Economics, which connects Post-Crash with radical student groups at other universities across the world. Because, as the story of Post-Crash shows, liberal education is not an institutional settlement but an ongoing collective struggle against closure. A struggle from which we can learn.

Notes

1 John Henry Newman, *The Idea of a University* (London: Longmans, Green, 1852).
2 Susan Wallace (ed.), *A Dictionary of Education*, 2nd edn (Oxford: Oxford University Press, 2015), doi: 10.1093/acref/9780199679393.001.0001.

3 Martha Nussbaum, 'Liberal education and global community', *Liberal Education*, 90:1 (2004), 42–7, https://files.eric.ed.gov/fulltext/EJ728534.pdf (accessed 28 March 2024).

4 James C. Scott, *Seeing Like a State: How Certain Schemes to Improve the Human Condition Have Failed* (New Haven, CT: Yale University Press, 1998).

5 Simon Kuper, *Chums: How a Tiny Caste of Oxford Tories Took over the UK* (London: Profile, 2022).

6 Joe Earle, Cahal Moran and Zach Ward-Perkins, *The Econocracy* (Manchester: Manchester University Press, 2017). See also the viewpoint of another founder, Maeve Cohen, 'Post-crash economics: have we learnt nothing?', *Nature*, 561 (13 September 2018), 151.

7 Richard Whitley, 'Changing governance of the public sciences: the consequences of establishing research evaluation systems for knowledge production in different countries and scientific fields', in Richard Whitley and Jochen Gläser (eds), *The Changing Governance of the Sciences: The Advent of Research Evaluation Systems* (Dordrecht: Springer, 2007), pp. 3–27; Frederic S. Lee, Xuan Pham and Gyun Gu, 'The UK Research Assessment Exercise and the narrowing of UK economics', *Cambridge Journal of Economics*, 37:4 (2013), 693–717, doi: 10.1093/cje/bet031.

8 Frederic S. Lee, 'The Research Assessment Exercise, the state and the dominance of mainstream economics in British universities', *Cambridge Journal of Economics*, 31:2 (2007), 309–25, doi: 10.1093/cje/bel021; Engelbert Stockhammer, Quirin Dammerer and Sukriti Kapur, 'The Research Excellence Framework 2014, journal ratings and the marginalisation of heterodox economics', *Cambridge Journal of Economics*, 45:1 (2021), 243–69, doi: 10.1093/cje/beaa054.

9 Earle et al., *The Econocracy*, p. 108.

10 Ibid., p. 106.

11 Ibid., p. 109.

12 Stockhammer et al., 'The Research Excellence Framework 2014', 245.

13 The University of Manchester Post-Crash Economics Society, *Economics, Education and Unlearning: Economics Education at the University of Manchester*, with a foreword by Andrew Haldane, https://www.rethinkeconomics.org/wp-content/uploads/2023/10/Post-Crash_Manchester-Curriculum-Review.pdf (accessed 26 February 2024).

14 Earle et al., *The Econocracy*, p. 108.

15 Ibid., p. 109.

16 Ibid., p. 107; 'Economics students aim to tear up free-market syllabus', *Guardian*, 24 October 2013, https://www.theguardian.com/business/2013/oct/24/students-post-crash-economics (accessed 28 March 2024).

17 Wendy Carlin Wikipedia page, https://en.wikipedia.org/wiki/Wendy_Carlin (accessed 30 October 2023).

Merger, global ambition and a renewed civic role – the University of Manchester from 2004 to 2024

Luke Georghiou

On 1 October 2004 the Victoria University of Manchester (VUM) and the University of Manchester Institute of Science and Technology (UMIST) combined to form the new University of Manchester. Belying its status as an exemplar for university mergers, regularly cited by both supporters and opponents of mooted restructures around the world, the coming together was not described as such in its planning and execution. In formal terms it was instead the double dissolution of the predecessor institutions and the foundation by Royal Charter of the new University of Manchester. This distinction is important in understanding the motives for this radical step and the subsequent development of the University during the final two decades of its bicentenary. Mergers of universities are typically associated with rationalisation, cost saving, achieving critical mass and *in extremis* the rescue of a failing institution. While elements of all but the last of these were present in the Manchester rationale, they were subordinate to an understanding that a changing world required new approaches and that these could only be achieved through the consensual design of a new university.

In this chapter we shall see how the initial rationale was embodied in the University and subsequently realised and elaborated in its first twenty years through long-term plans and ambitions and their interaction with events. This vicennial, coinciding with the bicentenary, has seen structures established at merger go on to be changed, sometimes radically. Iconic individuals have emerged, arrived and sometimes departed. The estate is transformed. A mixed legacy of heritage buildings and 1960s expansionism dominated by arterial roads has been enhanced by the addition of striking buildings and a setting in green spaces. Few in the 1970s would have imagined that the campus would feature as a leading tourist attraction for the city. As with most research universities, the student population has grown and become more internationalised, bringing both benefits and new challenges. One generation had their experience of university education radically disrupted by the COVID-19 pandemic. Other external seismic events affecting the University have included austerity, its consequences for inequalities with the emergence of 'levelling up' of regions as a policy driver, Brexit, and the climate crisis. This account will seek to show how the University's compass of social responsibility along with its heritage of innovation and 'Manchesterness' have guided it through the two decades.

The 2004 'merger'

The formation of the new University of Manchester was the culmination of a long history of association between VUM and UMIST.[1] Both had antecedents founded in 1824: for UMIST the Manchester Mechanics' Institution for the education of working men, and for VUM what was to become the Royal School of Medicine and Surgery, subsequently incorporated into Owens College in 1872. Owens College in turn became England's first civic university, the Victoria University of Manchester, receiving its Royal Charter in 1880. In the meantime, the Mechanics' Institution had evolved into a Technical School in 1883 and acquired the status of the Faculty of Technology of VUM in 1905, whereby students could be awarded the latter's degrees. This status persisted until 1955 when UMIST (at that time called the Manchester College of Science and Technology) received its own charter and direct national funding from the University Grants Committee. In 1994 UMIST became completely independent and began to award its own degrees. Cooperation between the two Manchester institutions was to be on the basis of an equal partnership.

It is with a degree of irony that the culmination of half a century of divergence formed the basis for the process of assuming a single identity. In the early 2000s both universities were considering their future strategies. Despite their clear strengths, the external environment posed both threats and opportunities. Competition was one such driver. To keep at the forefront of research was increasingly costly. Research challenges regularly demanded broadly based interdisciplinary teams and state-of-the-art facilities. In the sphere of education, competition was principally manifested in an increasingly globalised market for international students. UMIST was particularly dependent upon these, and with a narrower base of mainly high-cost subjects was vulnerable to shocks that might disrupt recruitment. Beyond the perceived benefits of scale and scope, the leadership of the universities also recognised that modernised governance and management structures would confer advantages of agility and efficiency in this uncertain environment.

Opportunity also lay in the peculiar economic geography of the United Kingdom, in which London and the south-east dominated not only the economy but also R&D expenditure and more specifically funding for university research. Many recognised that the future of the national economy and social fabric required the nation's other cities to close the gap in productivity and wider economic performance, and thus better the health and economic prospects of those who lived there and in the surrounding regions. Greater Manchester was seen by many, and certainly by its own leaders, as the best opportunity to escape the trap of a unimodal economy, and already possessed a global brand associated with post-industrial regeneration, albeit one resting largely upon its cultural and sporting assets. The identification of the need for an engaged world-class academic institution, matching but not emulating the Golden Triangle of Oxford, Cambridge and London, was a natural consequence of this line of thought.

Against this background a Joint Working Group was appointed to be independently chaired by Dr John Beacham, a regionally respected figure formerly of ICI (Imperial Chemical Industries). It was mandated by the two universities 'to consider various ways to develop a closer relationship between UMIST and the Victoria University of Manchester in order to build on existing strengths, with the aim of achieving world class standing in research, scholarship and teaching across a broad range of disciplines'. The group considered a range of options for cooperation between the institutions, ranging from encouragement of academics to undertake *grassroots collaboration* across separate institutions, through a *federal collaboration* for facilitating collaboration between separate institutions, a *trade model* whereby duplicated departments would be combined and assigned to one or the other, and a *combined institution*. To the surprise of some, it was this last option that was preferred by the group. Among the other options they saw high transaction costs and ultimately an insufficient base in either university to fulfil the ambitions being set for it. Combination would be achieved by the dissolution of the two existing universities to allow the creation of a single new chartered university. The desired step-change in performance would be achieved through enhanced research performance, increased student choice through new and interdisciplinary courses and a clear mission to enhance the economic well-being of the region.

Between April and October 2002, a process of due diligence on this recommendation was undertaken at the behest of the two governing bodies. Known as Project Unity, it involved wide consultation with stakeholders, including staff and students, and expert advice on issues such as finance and pensions. This was the design phase of the new University. A guiding principle was that no policy, procedure or structure would be adopted by default from what were now being perceived as the 'legacy institutions'. The process demanded review at all levels of activity, including governance, and the design of professional and support services including finance, human resources, estates and information technology. At the core was the academic design, including schools, faculties and the degree programmes to be offered. While no activity was left untouched, the focus of this effort was in those subjects that had been taught or researched in both universities. A need for colocation ensued with major consequences for the estates strategy, discussed later in this chapter. In turn this created financial demands, estimated at £300 million. The greater part of the investment came from redirecting formula allocations due to the universities from the Higher Education Funding Council for England. This was supplemented by developmental funds from the same body, research funding towards new facilities in photon science and neuroscience, regional support, and from the reserves and sale of surplus assets of the universities.

Following the due diligence, general agreement was secured to proceed. A timetable was set for full integration by 1 October 2004. It would be misleading to say that the process was entirely smooth. Staff concerns about potential job losses were addressed by a two-year moratorium from the date of foundation, though for a small, targeted group an early retirement and voluntary severance (ERVS) package was made available. The consequence of this guarantee surfaced in 2006 when the University was facing a substantial operating deficit, of which £10 million was partly attributed to this moratorium. Staff numbers had increased by 2,800 in thirty months, some of which was driven by research income and investment; nonetheless recovery entailed a tight grip on finance and a much larger ERVS scheme under which 630 people left the University. The other barrier to progress was a concern among some members of the governing body of UMIST that the distinctive heritage of the institution would be lost and its mission diluted in a full-service university.

The smoothness of the transition is a tribute to the two vice-chancellors, Martin Harris and John Garside, and the executive teams who supported them. Furthermore, the fact that both intended to retire removed the potential impediment of choosing one to lead the new University. Instead, a search was instigated for a 'president and vice-chancellor' (a title thought to have more international currency). The vice-chancellor of the University of Melbourne, Alan Gilbert, was selected to be the inaugural leader. He had an eight-month period as president and vice-chancellor designate, during which he had 'several months free from the burden of day-to-day operational management in which to build a senior leadership team, develop the "step change" agenda in consultation with colleagues across the merging institutions, and design the kinds of governance and management structures that the embryonic institution would need to facilitate its ambitions'.[2] This purposive approach to strategy has been a characteristic feature of the University in the two decades since. Gilbert's strategic plan was published in a document called 'Towards Manchester 2015'. After his retirement and subsequent untimely death in 2010, his former deputy and successor Dame Nancy Rothwell first refreshed the agenda in 2011 and then super-seded it with a strategic plan known as Manchester 2020. The final years of the vicennial have been guided by a vision and strategic plan called 'Our Future'. A common thread of ambition runs through these plans which has kept the University on a path consistent with the expectations associated with the merger.

Evolutionary change and reaction to changing circumstances has of course been a constant presence. For example, the initial academic structure of the University had been seen as a radical attempt to break down barriers through the creation of 23 large schools, several of which were interdisciplinary, and to manage these through four large faculties, each led by a 'vice-president and dean'. The dual title represented the expectation that the office holders would also share collective responsibility for University leadership (alongside four 'policy vice-presidents') as well as representing the focused interests of their faculties. Over time a logic of rationalisation prevailed, with the smallest of the faculties, Life Sciences, being merged with Medical and Human Sciences to form the Faculty of Biology, Medicine and Health, consisting of three schools correspond-ing to its titular components. A key part of the rationale was to offer a clear pathway from basic biology to application in health. In the other faculties rationalisation was at school level, with a series of mergers and reallocations of staff groups reducing the number to nine, all of which spanned broad interdisciplinary but cognate remits. Though hard to measure, external feedback has confirmed an internal sense that comparative ease of interdisciplinary working is a feature of the University. The merger had envisaged economies of scale, but economies of scope have arguably proved more important in that the University can internalise the combina-tions needed to address societal challenges and industrial needs.

An overlay of research institutes and 'platforms' have further helped to marshal expertise in key areas such as Digital and Sustainable Futures and Creative Manchester.

Manchester people: iconic and public figures

An early effort, championed by Alan Gilbert, was to accelerate the international standing of the University through the appointment of 'iconic scholars' whose virtuosity had been recognised internationally, encompassing but extending beyond the academic community. The opportunity to build on the already prominent record of Manchester (and its alumni) in securing Nobel Prizes provided an obvious pathway to meet this aim. A KPI of having at least five Nobel Laureates (or their equivalent in other fields) on the staff by 2015 was met with some scepticism in wider circles, but these doubts were quelled when Economics Laureate Joseph Stiglitz accepted a part-time role as chair of the Brooks World Poverty Institute in 2005. His appointment was followed in 2007 by that of Sir John Sulston, Laureate in Physiology or Medicine. Sulston was founding director of the Wellcome Trust's Sanger Institute and had led the British side of the Human Genome project. Like Stiglitz a man of high social conscience, Sulston came to Manchester not to pursue further his outstanding career in life sciences, but rather to work with bioethicist John Harris in what was to become the Institute for Science, Ethics and Innovation. His focus was on protecting the public interest, ensuring open access to scientific information and keeping commercial exploitation under control.

Sadly, Alan Gilbert's tragically early death meant that he did not live to see Manchester gain two further 'home-grown' Nobel Laureates when the 2010 Prize in Physics was awarded to Andre Geim and Konstantin (Kostya) Novoselov 'for groundbreaking experiments regarding the two-dimensional material graphene'. Their key publication in the journal *Science*, now widely cited, had been published in 2004, coinciding with the new foundation. The story, now legendary, involved peeling very thin layers off graphite using Scotch tape and ultimately isolating a single layer of atoms and discovering its extraordinary properties. This was a defining moment for the new University. It provoked press comment on the lines of an article in the *Independent* entitled 'Manchester: Britain's Greatest University' which stated: 'For the first time in living memory, the provincial red brick university has more Nobel Prize winners on its staff than either Oxford – which has none – or Cambridge, which has two.'[3]

Nobel Prizes address only certain subjects and it was always the intention of the University to bring in figures of equivalent standing from other fields. The original iconics cohort included a visiting professorship for the celebrated political scientist Robert Putnam, known for his work on declining trust and social capital, epitomised in the title of his book *Bowling Alone*.[4] The leading novelist Martin Amis was appointed as Professor of Creative Writing, one of several distinguished literary figures to teach in the University. Another was Professor of New Writing, the Manchester-born prize-winning novelist Jeanette Winterson. Her Foundation Day lecture in 2014 began by citing how the Manchester egalitarian spirit was built into the founding principles of the nineteenth-century institutions that became the University of Manchester. She went on to conclude that in this extraordinary city this extraordinary university should lead the country through its values and vision. In 2022 Lemn Sissay was appointed Honorary Professor of Creative Writing after completing his term as chancellor, which for many will be remembered for his poems capturing the spirit of the University, notably 'Making a Difference', which highlighted its commitment to social responsibility, and 'The World Wakes', celebrating graphene: 'The future – it's *you* in flatland'.

The University is also known for its academics who have become celebrated media personalities, including science and engineering popularisers Danielle 'Dan' George and Brian Cox (see Chapter 15). Prominence in the media is not confined to those in STEM subjects. Historian Michael Wood has produced many best-selling books and notable television series. In 2019 two significant public intellectuals joined the University: well-known writer and filmmaker David Olusoga, like Wood as Professor of Public History, and award-winning author, broadcaster and columnist Gary Younge, as Professor of Sociology.

Transforming and unifying the campus

At the time of the merger a traveller passing down Oxford Road or Sackville Street would have witnessed two campuses which had not fundamentally changed since the expansionary days of the 1960s and 1970s. Access to the Victoria University of Manchester from the north passed underneath the Precinct Centre, a complex built between 1970 and 1972. Its main elements were the Manchester Business School on Booth Street West, Crawford House to the east of Oxford Road, and between these buildings a desultory shopping centre rising above which gave the area its name. This was the vestige of a concept developed by the planners Hugh Wilson and Lewis Womersley in the 1960s which aimed to separate pedestrians from traffic by connecting buildings via elevated walkways. On the east side the Precinct Centre did indeed join up with the upper floor of what was then called the Computer Building, and on to a first-floor access to the Mathematics Building, an 18-storey tower. The scheme was never completed, meaning that on both sides of the road access from the south was via inconvenient ramps. Another high-rise building, the Moberly Tower, a postgraduate residence, sat above the Refectory between the Whitworth Hall and the Students Union building. The Medical School was housed in the Stopford Building from 1972, the largest building in the University and in its early years described as the most up-to-date.

On the UMIST campus the Main Building (later renamed the Sackville Street Building), the original part of which had been opened by Prime Minister Arthur Balfour in 1902, was the dominant edifice and was in its day a state-of-the-art facility for engineering education, described in concept as 'the greatest technical school at present existing in England'.[5] Most of the campus buildings dated from the expansion during the 1960s. The preponderant material was concrete. Of these, the most architecturally meritorious in the opinion of many was the Brutalist-style Renold Building, primarily a purpose-built

17.1 Manchester Municipal Technical School (Sackville Street Building), opened in 1902. Courtesy of the University of Manchester.

facility for lecture theatres and exhibitions. The campus uniquely featured in close proximity a motorway (the elevated section of the Mancunian Way), a railway line, a canal and a river (the Medlock), which runs through underground culverts.

Transformation of the campus was a key feature of the merger. Alumni passing down the same streets today sometimes struggle to orient themselves among the many new buildings set in parkland reclaimed from built-up areas and in one case replacing the busy connecting route, Brunswick Street. Much of the main artery, Oxford Road, has changed from being a source of pollution and a constant hazard of speeding traffic to a pedestrian and cycle-friendly environment with broad tree-lined pavements and traffic restricted almost exclusively to public transport.

The changing campus was driven by a series of masterplans. The first of these, the 2004 Estates Strategy and Masterplan, was the result of preparations for the merger and sought to create an urban university integrated with the surrounding city. In itself this was a departure, as the starting point was a campus bounded to the east and west by walls and buildings that deterred access by the local community. To meet the needs of the merged University the aim was to create a cohesive campus of improved functional buildings and surrounds which facilitated the academic aims of co-locating duplicate cognitive units while advancing interdisciplinary synergies.

This was to be the largest capital construction programme in the history of British higher education. The cost was offset by the disposal of older buildings, mainly residential, to reduce the size of the eventual estate and to defray the newbuild costs. There were 27 separate projects ranging from leading-edge research facilities such as the John Garside Building housing the Manchester Interdisciplinary Biosciences Centre, through several buildings named after Nobel Laureates and designed to house schools, to the mundane but necessary multistorey car park and infrastructural works. The largest edifice was University Place, a central hub with catering facilities, classrooms and the campus's largest lecture theatres, housed in a drum-shaped extension that dominates Oxford Road in the way that the demolished Mathematics Tower had on the same site. To the credit of the University's Estates and Services Directorate, the overall programme was completed on time and within budget, and both the overall strategy and individual elements garnered numerous professional awards. The merged campus appeared on maps as a skewed figure-of-eight, with the former UMIST campus closest to the city centre referred to as the North Campus, and the larger concentration to the south of the Mancunian Way.

A decision ratified by the board of governors in 2012 was to mark a watershed for the campus estate and in some respects the logical completion of the merger. The Campus Masterplan 2012–2022 announced the intention to create a single campus. In the words of the director of Estates and Facilities Diana Hampson: 'Since the merger of the two universities in 2004, it has been our ambition to bring all of the academic activity together on a single site south of the Mancunian Way, which will improve efficiency, improve the student experience and reduce the University's carbon footprint.' This would to entail the relocation of all academic accommodation from the North Campus to a site immediately adjacent to the South Campus, on Booth Street East, occupied at that time by the Grosvenor Halls of Residence and the Materials Science Building.

The ambitious replacement, initially called the Manchester Engineering Campus Development or MECD, and subsequently renamed in honour of Nancy Rothwell on her retirement, would accommodate most of the University's engineering teaching and research and was to be the single largest construction project ever delivered at the University. It was funded primarily through a public bond. In an entirely new experience for the University, the first necessary step was to gain a rating from the risk assessment agency Moody's, which to some relief was Aa1 Stable. The bond issue was four times over-subscribed, raising £300 million on a forty-year basis. The bonds are listed on the London Stock Exchange.

Handed over by the construction team in May 2021, the complex of buildings, including the eight-storey, 80,000 sq m Engineering Building A, provided a step-change in how a population of some 8,000 students would study and learn (plate 9). Long street-like spaces provided comfortable places to work and interact, while glass walls offered a view of research and teaching laboratories at work. In an unplanned step towards interdisciplinarity, the facility proved very attractive to students from other faculties, and projects spanning engineering, medicine and humanities took shape, reflecting conversations struck up while working adjacently.

A further highlight of the 2012 Masterplan was the transformation of the Business School. Following the merger, the school had combined four previously separate elements, the original Manchester Business School, UMIST's Manchester School of Management and, from the Victoria University, the Department of Accounting and Finance and the Manchester Institute of Innovation Research. In a rare bottom-up cross-institution initiative, these departments had already worked closely together under the banner of the Federal School of Business and Management. Nonetheless, a legacy of scattered and sometimes unfit for purpose buildings had been a barrier to further integration. With the support of a record £15 million donation from Lord Alliance, competitively secured national funding and a substantial University investment, a plan was implemented to convert the whole Precinct Centre into a breathtaking new home for the renamed Alliance Manchester Business School. The bridge over Oxford Road was demolished in August 2017, recorded in a viral 50-second time-lapse video. Old-timers remembered the surreal view from passing buses of whirling couples in the dance school that had occupied the suite that spanned the road.

The construction story of the campus is also a mirror of the University's research progress, most clearly visible in the area of advanced materials but also evident in other domains such as cancer research. Following the award of the Nobel Prize for graphene, University leaders were determined that this would not be another case of the cliché whereby a breakthrough in the United Kingdom is exploited elsewhere. With support from the city leaders a direct appeal was made to the centre of government, and the interest of the Chancellor of the Exchequer, George Osborne, was engaged. Funding for a world-class research facility, the National Graphene Institute (NGI), was made available outside the usual channels, supplemented by European Union Structural Funds. The NGI has 1,500 sq m of top specification cleanrooms featuring an atmosphere more than a million times purer than air. It also marked a step forward in the architecture of research buildings, including substantial input from Laureate Kostya Novoselov.[6] The graphene momentum continued when the opportunity emerged for a large investment from Abu Dhabi on the back of relationships built with the city following the acquisition of Manchester City Football Club. Matched by a coalition of UK public funders, the £60 million Masdar Building was erected to house the Graphene Engineering Innovation Centre or GEIC (pronounced geek from its conception). This was created to bridge the gap with business. The materials story continued when Osborne, seeking to address the national imbalance of R&D investment, challenged northern universities at a meeting in Manchester in June 2014 to suggest a 'Crick of the north' to equal London's Francis Crick Institute for biomedical science. By the next

Autumn Statement he was announcing £235 million towards the Sir Henry Royce Institute for Advanced Materials Research, a national institution spread across leading national universities but with its hub being a major building and facility hosted by the University.

A socially responsible university

From the time of the merger, the agenda for 2015 had included two goals with a social element, the first setting a commitment to 'Widening Participation' by making Manchester the UK's most accessible research-intensive university. A key manifestation of this goal was the devotion of a higher proportion of the University's fee income than any of its peers to support merit-based scholarships for both UK and international students from disadvantaged backgrounds. There was also a goal committing Manchester to provide 'More Effective Service to the Community'. The updated strategic plan in 2011/12 elevated social responsibility to the status of a third strategic priority, alongside world-class research and outstanding learning and student experience. This remains a unique feature among universities in the UK, and is for many a key distinctive feature of the University. Social responsibility encompasses a portfolio of activities including public and civic engagement, equality and diversity, environmental sustainability, cultural engagement, volunteering by staff, students and alumni and widening participation. It also forms a unifying theme running through research, teaching and operational processes. It is fair to say that social responsibility underpins an area of shared ethos among staff and students and maintains the tradition founded in being Britain's first civic university. At a practical level many actions express this commitment; for example, the School Governors Initiative, which has helped over 1,000 staff and alumni to find volunteering placements as governors in local schools, or the work of the Humanitarian and Conflict Response Institute which studies and supports global health, international disaster management and peacebuilding. It has close links to UK-Med, a charity founded by emergency medicine specialist Tony Redmond, which provides the UK's only official emergency medical team. Based on the campus it regularly deploys a full field hospital to natural disasters or war zones to care for local communities.

Social responsibility activity has won several accolades for the University, including its being the only university to feature in the world's top ten in each annual iteration of the Times Higher Education Impact Rankings based on contribution to the United Nations Sustainable Development Goals, coming first in the world in 2021 and second in 2023. Not all aspects of the social responsibility agenda are a tale of success. Along with most of its national peers and other organisations in Manchester, the University is lagging in progress towards its commitment to reach zero carbon by the city's target year of 2038, with affordability the most visible barrier. Highlighting social responsibility also invites responses from sections of the community affected by the University's presence, notably in the context of plans for expanded student accommodation, whether or not they involve the University.

Awards have also gone to the University's cultural institutions, which form a key element of its public engagement activity as well as supporting other goals. The Whitworth Art Gallery, after a £15 million development in 2015 which blended the nineteenth-century building with the adjoining park, won architectural awards from RIBA and the Art Fund Prize for Museum of the Year. Not to be outdone, the University's celebrated radio astronomy facility situated in the Cheshire countryside at Jodrell Bank was designated a UNESCO World Heritage Site in July 2019. It was the first such in the UK to be named for its pioneering science. The adjoining Discovery Centre also benefited, with support from the National Lottery Heritage Fund and other donors to build the First Light Pavilion, an innovative building blending with its surroundings and housing a planetarium and a permanent exhibition of the site's history. The 130-year-old Manchester Museum also underwent a transformation, reopening its doors in 2023 with a narrative built around social responsibility, taking a leading role in the repatriation of sacred objects to their communities of origin. New galleries enhanced the wide-ranging original collections covering the heritage of South Asia, connecting to the diaspora in the city, and deepening the understanding of Chinese culture and its long-standing links with Manchester.

Bumps in the road

The pathway over twenty years has not always been smooth. External events and internal stresses and strains have also had to be faced. An input/output analysis of the experience of being a student at Manchester would highlight the University's consistent position as the nation's most popular university for undergraduate applications, and the most targeted by the UK's top graduate employers, and supported by the largest community of alumni whose collective achievements are quite mind-boggling. Nonetheless, the period between arrival and departure has sometimes proved more challenging, with scores in the National Student Survey below what the University aspires to. Cultural observer and former chair of the University's Alumni Association Andy Spinoza notes in his autobiographical account of the city of Manchester the 'student radicals whose tradition of protest stubbornly runs through its history like a trace of DNA'.[7] This is as much a facet of 'Manchesterness' as the self-reliance, solidarity and can-do attitude that pervade the city's culture.

The COVID-19 pandemic forced all universities to pivot to online provision and remote working in a matter of days. Distancing requirements and closed entertainment facilities challenged the nature of student life. Manchester had a longer and more severe experience of lockdown restrictions than almost anywhere in the UK. Many students nonetheless preferred to remain in their residences, notably at the Fallowfield complex. With all other entertainment in the city closed, the site became a magnet not only for other young people unconnected to the University but also for criminal elements seeking to exploit the situation. In an abortive attempt to exclude these groups, University services sought to seal off unauthorised access to the site by erecting a temporary perimeter fence. This elicited a protest involving some 1,000 students at its peak, dubbed by the press as 'Fencegate', and culminated in an apology from the University for distress caused. A subsequent inquiry highlighted failings in process and decision making including lack of communication and mission creep.[8] A postscript on the pandemic should acknowledge that the careful implementation of safety measures meant that no deaths resulted from COVID contracted on campus. Through medical provision, research and volunteering, staff and students contributed enormously to the fight against the virus.

The latter years of the post-merger period have also been subject to waves of industrial action mainly by members of the University and College Union. Most have been nationally driven with disputes around changes to the Universities Superannuation Scheme and pay claims. The local climate was affected in 2017/18 when as part of an initiative to create investment capacity to improve research and student experience and to achieve financial sustainability in fulfilment of the Manchester 2020 Strategic Plan, the University entered consultation with the trade unions in relation to reductions of up to 171 posts and with the possibility of compulsory redundancy if numbers were not achieved through ERVS. Irrespective of the rationale, the collegial nature of the University was strained by this episode.

Innovation and the future

The size of the University, its two hundred years of history, international profile, research achievements, alumni, and association with a city that many see as both the UK's most culturally dynamic and as one of the most beset by health and economic inequalities mean that it is not possible to do justice to the full richness of events, achievements and challenges that have characterised the twenty years since the merger. It is not appropriate for an insider to pronounce judgement on whether the expectations and goals of 2004 have been fulfilled, but perhaps that is the wrong question. In a constantly changing world, with geopolitical instability, economic crashes, the climate crisis and populist challenges to the values represented by most universities, progress demands constant evolution and adaptation of strategies and tactics to meet educational, research and innovation needs. In its bicentenary year it is fitting that the most strategic project being undertaken by the University is the conversion of the former UMIST campus into a £1.7 billion innovation district at the heart of the city. The University has entered a joint venture with developers Bruntwood Scitech, seeking to bring together business, education and communities (plate 10). For the University the principal reward lies less in property development and more in the intangible benefits that will come from a site in close proximity populated by firms, from large corporates to spinouts, that will be research collaborators, employers and placement hosts for students, and that will magnify the University's already considerable innovation ecosystem. As new president and vice-chancellor Duncan Ivison takes office after fourteen years of inspiring leadership by Nancy Rothwell, the transformation of this site can be seen as the last act of the merger, and with a certain symmetry as a reaffirmation of the roots of the University two centuries ago.

Notes

1 This section is a summary of a more detailed account given in Luke Georghiou, 'Strategy to join the elite: merger and the 2015 agenda at the University of Manchester – an update', in Adrian Curaj, Luke Georghiou, Jennifer Cassingena Harper and Eva Egron-Polak (eds), *Mergers and Alliances in Higher Education: International Practice and Emerging Opportunities* (Heidelberg: Springer, 2015), pp. 205–20.

2 Alan Gilbert, 'The Manchester merger, stretching the golden triangle', in Hugo de Burgh, Anna Fazackerley and Jeremy Black (eds), *Can the Prizes Still Glitter? The Future of British Universities* (Buckingham: University of Buckingham Press, 2007).

3 Jonathan Brown, 'Manchester: Britain's greatest university?', *Independent*, 9 October 2010, p. 18, www.independent.co.uk/news/education/education-news/manchester-britain-s-greatest-university-2101828.html (accesssed 28 March 2024).

4 Robert Putnam, *Bowling Alone* (London: Simon and Schuster, 2001).

5 H. E. Roscoe, 'The Manchester Municipal Technical School', *Nature*, 47 (1892), 201–4.

6 James H. Baker and James Tallentire, *Graphene: The Route to Commercialisation* (New York: Jenny Stanford Publishing, 2022); Kostya S. Novoselov and Albena Yaneva, *The New Architecture of Science: Learning from Graphene* (Singapore: World Scientific Publishing, 2020).

7 Andy Spinoza, *Manchester Unspun: Pop, Property and Power in the Original Modern City* (Manchester: Manchester University Press, 2023), p. 239.

8 University of Manchester Inquiry report: An investigation into the erection of fencing at the Fallowfield Halls of Residence on 5 November 2020 (University of Manchester, 2020), https://documents.manchester.ac.uk/display.aspx?DocID=52105 (accessed 5 January 2024).

Epilogue

Dame Nancy Rothwell

The University of Manchester's relationship with our city-region dates back two hundred years. The Manchester Mechanics' Institution was established by local businessmen to train people of the region in the skills needed for local jobs in the 'original modern' city. In the second half of the nineteenth century Owens College was deliberately fashioned to become England's first civic university. Since then, we have maintained a close relationship with our neighbours, and never more so than today, and our histories have been closely intertwined. Our relationship with our city-region has been a particular passion of mine and has been empowering to those within the University and in our region. We are proud to be in the home of the Industrial Revolution, of the trade union movement and of the suffragettes (a number of our staff having been arrested for protesting).

This partnership is ever evolving and is evident in many ways. For example, we are the most popular university for UK undergraduate student applications, likely due to a combination of the attractions of the city and our own attributes. In the extensive survey of staff, students and stakeholders for the development of 'Our Future', our vision and strategic plan launched in 2020, the word 'Manchesterness' featured strongly, though definitions vary.[1] Our location came out as a highly valued element, and civic engagement features as a key priority.

University core goals

For more than a decade, our three core goals have been research and discovery, teaching and learning, and social responsibility. More recently we have added three further, cross-cutting themes: innovation, global influence and civic engagement. Our explicit focus on social responsibility has made us unique among UK universities and has been recognised by our position in the Times Higher Education Impact Rankings, where we have featured among the top ten in the world since the ranking assessment began.[2] Each of these goals and themes has had an impact on, and has benefited from, engagement with leaders and neighbours in our region, and our strong partnerships. When asked how we would measure social responsibility there were no league tables, but I said that our neighbouring communities would talk of 'our university'.

E.1 University of Manchester Core Goals and Themes, from *Our Future: Vision and Strategic Plan 2020*.

Epilogue

Research excellence

Our research, which spans from basic discovery to applications, aims to be internationally recognised, and indeed in the most recent Research Excellence Framework in 2021, 93 per cent of our research was deemed internationally excellent or world-leading.[3] Much of our research is also highly relevant to our region and informs local policies and strategic priorities and ambitions, with our staff and postgraduate researchers working in close partnership with relevant organisations in Greater Manchester. Just a few examples of this are our Tyndall Centre, which advised Greater Manchester on future zero carbon targets; our Institute of Education, which partners with many local schools, and includes the Beewell project on well-being in school children; Creative Manchester, one of our cross-University platforms which works with cultural institutions across the region; and our research on poverty, inequalities and justice in the workplace.

The interdependence is particularly important to our research across many areas of health, and it is notable that our Medical School owes its origins to a school established two hundred years ago. Greater Manchester overall suffers from very poor health compared to the south-east of the UK and this has major economic, educational and social impacts. Our University-wide programme on health inequalities aims to identify interventions that can most effectively close the gap in health outcomes for all and with significant benefits to our city-region.

Education, skills and 'giving back'

Our student population is probably the largest on a single-site campus in the UK (almost 45,000) and includes almost 15,000 international students; 20 per cent of all students come to study with us from the north-west. Greater Manchester is not where it wants and needs to be in terms of the skills needed for employment and educational attainment, though this is steadily improving. Each year we recruit around 1,200 students from local areas, approximately 50 per cent of whom are from widening participation backgrounds, which means talented students from the most disadvantaged backgrounds.

Our Manchester Access Programme (MAP) has been instrumental in supporting local students who have the ability but perhaps not the support or aspiration to go into higher education. MAP is part of our wider access initiatives to support about 500 local students a year from less privileged backgrounds to access higher education. I am always so impressed by what they achieve.

I'm proud to say that almost half of our graduates stay in the region to study further or take up local employment. This has changed significantly over recent years, to the point where we are now a 'net importer' of graduates. While they are studying here, our students contribute to the local community in many ways, including to the economy. They deliver thousands of hours of volunteering each year, much of this in supporting younger students in their learning, for example through the Tutor Trust.

Commitment to social responsibility

Our activities under this goal are often undertaken locally. For example, our work in schools, helping long-term unemployed residents to gain valuable jobs, and our local community panels which come together to advise on what they would like to see from the University.

Public engagement, communicating to a wider audience, has long been an important activity for us, and we were the first university to be given a gold award from the National Centre for Community and Public Engagement; just recently we received a platinum award, the new highest category.

Our cultural institutions, the Manchester Museum, the Whitworth Art Gallery, Jodrell Bank Discovery Centre and the John Rylands Library, and many events within faculties attract hundreds of thousands of visitors each year, many of them local.

In planning our campus, we have been mindful of making it an open and welcoming area for our neighbours, with many more green spaces developed, particularly since 2004, and the development of Brunswick Park, which has been widely welcomed by our neighbours. We can now see many schoolchildren walking across the campus and popping into our buildings. In MECD they sit alongside our students studying and using the free wi-fi.

Driving innovation

Innovation, the translation of new ideas, products and processes into economic and social value, is certainly topical, but has long been one of our traditions. In 1944 the Victoria University of Manchester, one of our predecessor institutions, established a 'Manchester Joint Research Council' with the Manchester Chamber of Commerce. Its objectives were to bring science and industry closer, and particularly to help smaller firms; to spread new knowledge, and to accelerate its use and application; to encourage research; and to initiate discussions, pursue special interest inquiries and investigations on industrial, scientific and related subjects, including the consideration of economic and sociological problems.

We are pursuing the same aims today, though of course now innovation is much broader than sciences. Many of the companies formed from discoveries in the University, largely through our Innovation Factory, remain locally. Thousands of students who take courses as part of the Masood Enterprise Centre in our Alliance Manchester Business School go on to start or support companies, and Northern Gritstone, a venture capital fund developed jointly by us and the universities of Sheffield and Leeds, is providing essential start-up funding for new enterprises.

By far the largest opportunity in innovation is through our £1.7 billion Innovation District Manchester on the former UMIST (North Campus) site, which should create over 10,000 new jobs.[4] This provides an excellent example of our ever evolving partnership with our city-region.

The best of local and global

Manchester has always been a global city with a very diverse population, much international trade and a welcoming place to people from across the world. Similarly, the University, with many international staff and students, global partnerships with universities and companies, and centres in Hong Kong, Shanghai, Singapore and Dubai, very much sees itself on a global stage and uses international measures of success. We partner with industry and regional government in the Manchester China Forum and the Manchester India Forum, we work closely with Manchester Airport, particularly to facilitate international student travel and arrivals, and we frequently share knowledge and itineraries for numerous international visitors.

We have at least 550,000 alumni and over a quarter of these live outside the UK. They are extremely valuable to us in forging global links, and attracting international visitors and students, investors and companies. This global reach complements our local priorities.

Partnering for success

Greater Manchester as a city-region has a strong history and many achievements, and is recognised for the strength of its leadership over nearly three decades. In 2011 the Greater Manchester Combined Authority (GMCA) was established, comprising the ten local councils. Then, in 2017 the first elected regional mayor, Andy Burnham, took up post. Where Greater Manchester led, many other city-regions have now followed. The University has been proud to partner in, and benefit from, many of these achievements. My predecessor, Professor Alan Gilbert, recognised the importance of a strong relationship with our regional leaders, and we have built on this. We have benefited from very close relationships with past and current leaders, who all recognise the value of the University to our region, and we in turn value our close relationship with leaders. Independent estimates indicate that the University of Manchester adds £1.5 billion per annum to the regional economy and leads to the creation (directly and indirectly) to over 20,000 jobs in addition to our own 2,000 jobs.

Our university, Manchester Metropolitan University, the Royal Northern College of Music and the universities of Salford and Bolton work closely together on many issues. Since the COVID lockdown the five vice-chancellors have met regularly to share experiences, opportunities and challenges. Greater Manchester formed the first city-region-wide Civic University Agreement in 2021 between all five universities, which has already delivered success in driving social, economic and environmental change.[5] We also commissioned an independent survey of local residents to assess their views of our universities.[6] The outcome was more positive than we might have expected, but also highlighted areas where we could do more for our neighbouring communities. The Greater Manchester universities have worked together with local business leaders and staff in GMCA on bids for government innovation funding and are developing plans for collaboration on creative activities and on zero carbon.

In 2015 the health budget of about £6 billion per annum was devolved to Greater Manchester and, as a direct result of this, Health Innovation Manchester (HInM) was established to drive discoveries and innovation into the health service and social care. We, along with all the major hospitals, industry leaders and regional government, have been members of HInM since its inception and this has undoubtedly benefited not only the health of our population but also our own research. In 2022 we, together with our partner hospitals, received the biggest national uplift in funding from the National Institute for Health Research for the Manchester Biomedical Research Centre, £59 million awarded over five years.

More recently a similar partnership was established through Innovation Greater Manchester (IGM). This was instrumental in securing £33 million government funding as part of the Greater Manchester innovation accelerator, from which we and other regional universities benefited, with ID Manchester at its heart, but also including other Greater Manchester sites.

In our more immediate vicinity, the Oxford Road Corridor Board (ORC) includes ourselves, Manchester Metropolitan University, Bruntwood, Manchester City Council and Manchester University NHS Foundation Trust. ORC has delivered numerous economic, cultural, environmental and social benefits to the region around the University.

There are many other examples of where we work closely with regional partners, for example in skills, delivering zero carbon, transport, inward investment, support for local businesses, and of course in lobbying government for funding and other support for Greater Manchester.

Summary and future

There is no doubt that our location and the fact that we have been so embedded in Manchester and the wider city-region has shaped the University for two hundred years. Each has empowered and strengthened the other. It has been a privilege to know that I can pick up the phone (almost on speed dial) to our local leaders and will get a fast response.

In reflecting on my time as president and vice-chancellor, one of the areas I feel most proud of is the strength of our numerous relationships with our city-region. I have no doubt that this success will continue as we enter our third century.

Notes

1 'Our future: knowledge, wisdom, humanity', https://www.manchester.ac.uk/ discover/vision/ (accessed 26 February 2024).
2 'Number one in the UK and Europe for sustainable development', https:// www.manchester.ac.uk/discover/social-responsibility/sdgs/ (accessed 26 February 2024).
3 Research Excellence Framework 2021, https://www.manchester.ac.uk/research/ impact/ref-2021/ (accessed 26 February 2024).
4 https://www.id-manchester.com/ (accessed 26 February 2024).
5 'Civic University Agreement', https://www.manchester.ac.uk/discover/ social-responsibility/civic/civic-agreement/ (accessed 26 February 2024).
6 'The people's priorities for Greater Manchester's universities', https://www. publicfirst.co.uk/the-peoples-priorities-for-greater-manchesters-universities. html (accessed 26 February 2024).

Select bibliography

A book of this kind is necessarily eclectic in its source base, and bibliographical references on the subjects dealt with in particular chapters are most easily located in the endnotes to the chapters in question. This bibliography instead seeks to provide guidance on further reading on the two themes that run through the book: the history of the University of Manchester and its institutional forebears, and (much more selectively) aspects of the history of Manchester and its civic and cultural life.

University history

Cardwell, D. S. L. (ed.), *Artisan to Graduate: Essays to Commemorate the Foundation in 1824 of the Manchester Mechanics' Institution, now in 1974 the University of Manchester Institute of Science and Technology* (Manchester: Manchester University Press, 1974)

Charlton, H. B., *Portrait of a University 1851–1951, to commemorate the centenary of Manchester University* (Manchester: Manchester University Press, 1951)

Fiddes, Edward, *Chapters in the History of Owens College and of Manchester University, 1851–1914* (Manchester: Manchester University Press, 1937)

Fowler, Alan, and Wyke, T. J., *Many Arts, Many Skills: The Origins of the Manchester Metropolitan University* (Manchester: Manchester Metropolitan University, 1993)

Pullan, Brian, *A Portrait of the University of Manchester* (London: Third Millennium, 2007)

Pullan, Brian, with Abendstern, Michele, *A History of the University of Manchester 1951–73* (Manchester: Manchester University Press, 2000)

Pullan, Brian, with Abendstern, Michele, *A History of the University of Manchester 1973–90* (Manchester: Manchester University Press, 2004)

Robertson, Alex, and Lees, Colin, *The University of Manchester, 1918–1950: New Approaches and Changing Perspectives*, special issue of the *Bulletin of the John Rylands Library*, 84:1–2 (2002)

Select bibliography

Swinton, Jonathan, *Alan Turing's Manchester* (Manchester: Infang, 2019)

Thompson, Joseph, *The Owens College: Its Foundation and Growth; and Its Connection with the Victoria University, Manchester* (Manchester: J. E. Cornish, 1886)

Tylecote, Mabel, *The Education of Women at the University of Manchester, 1883 to 1933* (Manchester: Manchester University Press, 1941)

Whyte, William, *Redbrick: A Social and Architectural History of Britain's Civic Universities* (Oxford: Oxford University Press, 2015)

Manchester, past and present

Ayshford, John, et al., *The Simons of Manchester: How One Family Shaped a City and a Nation* (Manchester: Manchester University Press, 2024).

Crinson, Mark, *Shock City: Image and Architecture in Industrial Manchester* (London: Paul Mellon Centre for Studies in British Art, 2022)

Gunn, Simon, *The Public Culture of the Victorian Middle Class: Ritual and Authority in the English Industrial City, 1840–1914* (Manchester: Manchester University Press, 2000)

Howe, Anthony, *The Cotton Masters 1830–1860* (Oxford: Clarendon Press, 1984)

Hulme, Tom, *After the Shock City: Urban Culture and the Making of Modern Citizenship* (Woodbridge: Boydell, 2019)

Kidd, Alan, *Manchester* (Keele: Ryburn, 1993); 4th edition, *Manchester: A History* (Lancaster: Carnegie, 2006)

Kidd, Alan, and Wyke, Terry (eds), *Manchester: Making the Modern City* (Liverpool: Liverpool University Press, 2016)

Simon, Shena, *A Century of City Government: Manchester, 1838–1938* (London: Allen and Unwin, 1938)

Spinoza, Andy, *Manchester Unspun: Pop, Property and Power in the Original Modern City* (Manchester: Manchester University Press, 2023)

Wildman, Charlotte, *Urban Redevelopment and Modernity in Liverpool and Manchester, 1919–1939* (London: Bloomsbury, 2016)

Williams, Bill, *Jews and Other Foreigners: Manchester and the Rescue of the Victims of European Fascism* (Manchester: Manchester University Press, 2013)

Wolff, Janet, and Savage, Mike (eds), *Culture in Manchester: Institutions and Urban Change since 1850* (Manchester: Manchester University Press, 2013)

Index